MATERIALS AND PROCESSES FOR NDT TECHNOLOGY

The American Society for Nondestructive Testing

Materials and Processes for NDT Technology was edited by:
Harry D. Moore

Publication and review of this text was under the direction of the Personnel Training and Certification Committee of the American Society for Nondestructive Testing:

George Wheeler, Chair (1976-80)
Frank Sattler, Vice Chair (1976-79)
Robert Anderson, Secretary (1976-79)
F.C. Berry (1976-78)
Chet Robards (1976-79)
Carl Shaw (1976-82)
Kermit Skeie (1976-79)
Robert Spinetti (1976-80)

Allen Whiting (1976-1979)
Robert Brostrom (1978-81)
Ward Rummel (1978-81)
John Weiler (1978-81)
Robert Baker (untenured)
F.N. Moschini (untenured)
Ed Briggs (1979-82)
Jack Spanner (1979-82)

Contributors:
Donald R. Kibby
Dr. Robert C. McMaster
Dr. Vernon L. Stokes

ASNT Production Staff:
Robert Anderson, Technical Director
George Pherigo, Director of Education
Diana Nelson, Coordinator of Education Services

Published by
The American Society for Nondestructive Testing, Inc.
1711 Arlingate Lane
PO Box 28518
Columbus, OH 43228-0518

Copyright © 1981 by The American Society for Nondestructive Testing, Inc. ASNT is not responsible for the authenticity or accuracy of information herein. Published opinions and statements do not necessarily reflect the opinion of ASNT. Products or services that are advertised or mentioned do not carry the endorsement or recommendation of ASNT.

IRRSP, *NDT Handbook*, *The NDT Technician* and www.asnt.org are trademarks of The American Society for Nondestructive Testing, Inc. ACCP, ASNT, *Level III Study Guide*, *Materials Evaluation*, *Nondestructive Testing Handbook*, *Research in Nondestructive Evaluation* and *RNDE* are registered trademarks of The American Society for Nondestructive Testing, Inc.

ASNT exists to create a safer world by promoting the profession and technologies of nondestructive testing.

ISBN-13: 978-0-931403-06-4
ISBN-10: 0-931403-06-5

Printed in the United States of America

first printing 1981
second printing 12/87
third printing 05/88
fourth printing 10/89
fifth printing 12/91
sixth printing 01/94
seventh printing 04/96
eighth printing 08/98
ninth printing 11/00
10th printing 09/05
11th printing 10/07
12th printing 02/09
13th printing 04/11
14th printing 08/13
15th printing 11/14

Table of Contents

CHAPTER		Page
1	RELATION OF NDT TO MANUFACTURING	1

NONDESTRUCTIVE TESTING: NDT Definition. REQUIREMENTS FOR NDT SUPERVISORY PERSONNEL. QUALIFICATION AND CERTIFICATION OF NDT LEVEL III TESTING PERSONNEL: Sources of Technical Information Available to NDT Personnel. MANUFACTURING: MATERIALS AND PROCESSES: Material Failures. Purpose for Use of NDT. NDT IN FRACTURE CONTROL.

2	INTRODUCTION TO MANUFACTURING TECHNOLOGY	9

History. INDUSTRIAL RELATIONSHIPS: Competition in Industry. Personnel. Nomenclature. SUBJECT MATTER: Materials. Processes. Economics. Order.

3	PROPERTIES OF MATERIALS	15

INTRODUCTION: Classes of Properties. Significance of Properties of Design. LOADING SYSTEMS AND MATERIAL FAILURE: Loading Systems. TESTING: The Tensile Test. True Stress-True Strain. Compression Testing. Transverse Rupture Testing. Shear Testing. Fatigue Testing. Creep Testing. Notched Bar Testing. Bend Testing. Hardness Testing. Factor of Safety.

4 THE NATURE OF MATERIALS AND
SOLID STATE CHANGES IN METALS   31
The Effect of Energy on the Atom. Metallic Structure. Solidification. Grain Size. SOLID STATE CHANGES IN METALS: Work Hardening. Plastic Deformation. Cold Work. RECRYSTALLIZATION: Recovery. Recrystallization. Grain Growth. AGE HARDENING. ALLOTROPIC CHANGES. HEAT TREATMENT OF STEEL: Approximate Equilibrium Heat-Treatment Processes. Austenitization. Annealing. Normalizing. Spheroidizing. Hardening of Steel. Tempering. CORROSION: Direct Chemical Action. Electrolytic (Electrochemical) Reaction. Corrosion Rate Dependent on Several Factors. Types of Corrosion. Corrosion Protection.

5 FERROUS METALS   45
Choosing Metals and Alloys. Ferrous Raw Materials. CAST IRONS. STEEL: Wrought Iron. Steel Making. Plain Carbon Steel. Alloy Steels. Low Alloy Structural Steels. Low Alloy AISI Steels. Stainless Steels. Tool and Die Steels. Cast Steels. MATERIAL IDENTIFICATION SYSTEMS.

6 NONFERROUS METALS AND
PLASTICS   57
ALUMINUM ALLOYS: General Properties. Wrought Aluminum Alloys. Property Changes. Cast Aluminum Alloys. COPPER ALLOYS: General Properties. Brasses and Bronzes. NICKEL ALLOYS. MAGNESIUM ALLOYS. ZINC ALLOYS. SPECIAL GROUPS OF NONFERROUS ALLOYS: Heat- and Corrosion-Resistant Alloys. Other Nonferrous Metals. NON-METALS: Plastics. Plastic Materials. Types of Plastics. Characteristics of Plastics.

7 THE NATURE OF MANUFACTURING   73
MODERN MANUFACTURING: Markets. Design. Processing. States of Matter. Shape-Changing Processes. Summary.

8 THE CASTING PROCESS   79
The Process. SOLIDIFICATION OF METALS: Solidification. Shrinkage. POURING AND FEEDING CASTINGS: Casting Design. Pouring. The Gating System. Risers. Chills. FOUNDRY TECHNOLOGY. SAND MOLDING: Green Sand. Patterns. Flasks. Sand Compaction. Cores. Green Sand Advantages and Limitations. Dry Sand Molds. Floor and Pit Molds. Shell Molds. METAL MOLD AND SPECIAL PROCESSES: Permanent Mold Casting. Die Casting. Investment Casting. Plaster Mold Casting. Centrifugal Castings. Continuous Casting. MELTING EQUIPMENT: Cupola. Crucible Furnaces. Pot Furnaces. Reverberatory Furnaces. Electric Arc Furnaces. Induction Furnaces. FOUNDRY MECHANIZATION.

9 THE WELDING PROCESS   95
BONDS: Nature of Bonding. Fusion Bonding. Pressure Bonding. Flow Bonding. Cold Bonding. WELDING METALLURGY: Composition Efects. Effects on Grain Size and Structure. Effects of Welding on Properties. DISTORTIONS AND STRESSES.

10 WELDING PROCESSES AND DESIGN   105
HEAT FOR WELDING: Chemical Reactions. The Electric Arc. Welding Equipment and Procedures. Arc Welding Electrodes. Modification of Arc Welding for Special Purposes. Automatic Welding. Electric Resistance Heating. SPECIAL WELDING PROCESSES: Electron-Beam Welding. Plasma Arc. Ultrasonic Welding. Friction Welding. Electroslag Welding. Explosion Welding. Diffusion Welding. WELDING DESIGN: Joints. Design Considerations. Weldability. WELD DEFECTS: Fusion Welding. Dimensional Defects. Structural Discontinuities. Weld Metal and Base Metal Properties. Basic Symbols for NDT.

11 PLASTIC FLOW   121
EFFECTS OF DEFORMATION: Work Hardening and Recrystallization. Effects of Flow Rate. Direction Effects. Temperature and Loading Systems Effects. Grain Size. RELATIVE EFFECTS OF HOT AND COLD WORKING: Mechanical Properties. Finish and Accuracy. Process Requirements.

12 MILLWORK, FORGING, AND POWDER
METALLURGY   127
MILLWORK: Hot Rolling. Cold Finishing. Tube and Pipe Making. Extrusion. FORGING AND ALLIED OPERATIONS: NDT of Forgings. Open Die Forging. Closed Die Forging. Forging with Progressive Application of Pressure. Powder Metallurgy. Pressing. Sintering. Sizing and Postsintering Treatments. Application for Powdered Metal Products.

13 PRESSWORKING OF SHEET METAL   141
Shearing. Bending. Drawing. New Developments in Sheet Metal Forming.

14 MACHINING FUNDAMENTALS   147
The Machining Process. Chip Formation. Cutting Tool Materials. Abrasives. MACHINE TOOLS: Machinability. Finish. NUMERICAL CONTROL.

15   **MISCELLANEOUS PROCESSES**   157
PLASTIC PROCESSING: Compression Molding. Closed Die Molding. Casting. Extrusion. Reinforced Plastic Molding. Postforming. Design Considerations. ADHESIVE BONDING. COMPOSITES: Laminates. Mixtures. METAL REMOVAL PROCESSES: Electrical Discharge Machining. Electrochemical Machining. Other Possible Material Removal Methods. DEPOSITION PROCESSES: Electroforming. GROSS SEPARATION PROCESSES: Torch Cutting. Friction Sawing.

16   **SURFACE FINISHING**   171
CASE HARDENING OF STEELS: Carburizing. Flame Hardening. CLEANING: Choice of Cleaning Method. Liquid and Vapor Baths. Blasting. ABRASIVE BARREL FINISHING: Wire Brushing. Polishing. Buffing. Electropolishing. COATINGS: Preparation for Coatings. Paints, Varnishes, and Enamels. Lacquers. Organic Coating Application. Vitreous Enamels. Metallizing. Vacuum Metallizing. Hot Dip Plating. Electroplating. CHEMICAL CONVERSIONS: Anodizing. Chromate Coatings. Phosphate Coatings. Chemical Oxide Coatings.

17   **INSPECTION**   183
INSPECTION PROCEDURES: Organization of Inspection. Quantity of Inspection. Process Control Charts. PRINCIPLES OF MEASUREMENT: Dimensional References. Tolerances. Sources of Measurement Variation. Basis for Measurement. INSPECTION EQUIPMENT: Micrometer Caliper. Other Adjustable Tools. Indicating Gages and Comparators. Fixed Gages. Surface Finish. Surface Finish Measurement. Surface Specification.

Index   199

Preface

This book has been compiled as a reference and source of general information concerning manufacturing for use by personnel involved in designing, using, or evaluating nondestructive testing of products and structures. The text material has been kept as general as possible to still retain technical value but broad enough to include all phases of manufacturing industry and most of the materials used. The depth of treatment has intentionally been kept low in order that NDT personnel without a great amount of formal education might gain an interest and develop understanding of the material.

However, the techniques of NDT are *not* included in the coverage. It is intended that persons using this text be already informed regarding that subject or acquire the necessary knowledge from other more specialized sources. Wherever practical, though, throughout the book mention is made of certain NDT methods that might be suitable for the kinds of defects under consideration. Some of the limitations of the methods are also indicated in some applications. One of the essential needs for satisfactory use of NDT is recognition of its limitations; knowledge of the source of defects and the materials in which they are found is an aid in determining the validity of any test and its evaluation.

Appreciation is hereby expressed:

to Grid Incorporated for giving permission to use large portions of *Manufacturing: Materials and Processes* by Harry D. Moore and Donald R. Kibbey, 1975,

to Vernon L. Stokes for providing many of the illustrations used and copy which served as a guide for some of the text material,

to Robert T. Anderson, Technical Director, ASNT, and George L. Pherigo, Director of Education, ASNT, for their invaluable assistance in selecting and organizing the material and for supplying the majority of the NDT tie-ins scattered throughout the text, and

to Diana Nelson, Coordinator of Educational Services, ASNT, for her aid in editing the text material.

H. Don Moore
1979

Relation of NDT to Manufacturing 1

NONDESTRUCTIVE TESTING

Nondestructive testing is a fundamental and essential tool for control of quality of engineering materials, manufacturing processes, reliability of products in services, and maintenance of systems whose premature failure could be costly or disastrous.

NDT DEFINITION

Like most complex procedures, NDT is not definable by a few simple words. Nondestructive testing is normally interpreted to mean the use of physical methods for testing materials and products without harm to those materials and products. Many inspection procedures such as dimensional measurements, visual examination for completeness, functional tests, and others, although required in a manufacturing process, are not normally considered part of a NDT program.

Nondestructive Tests Are Always Indirect. It is frequently important to know a property or characteristic of a material or product which, if tested directly, would be destructive. Therefore it becomes necessary to perform a nondestructive test on some property or characteristic which can be related to that about which knowledge is desired. The test may be very simple in some cases, but in others may be complex and difficult. However, in *every* case, reliable correlation must be established between the desired property and the measured property (or properties).

Correlation May Be Costly And Difficult. Analysis to provide accurate knowlege of the relation between a testable quality and one which cannot be tested directly without destruction is likely to require a great amount of knowledge, skill, and background experience together with good judgement which, in a broad sense, can be described as an instinctive knowledge of the laws of statistical probability.

Decisions to accept or reject following a test result must be based on a thorough knowledge of materials and the properties, processes and their effect on properties, test techniques, design requirements, product applications, service conditions, and suitable life expectancy. Clearly this much knowledge is seldom located in a single individual, and group decisions or consultations may be necessary.

NDT correlation may require the cooperation between test supervisors, designers, metallurgists, manufacturing personnel, customer personnel, and test personnel.

REQUIREMENTS FOR NDT SUPERVISORY PERSONNEL

From the above it can be seen that supervisory personnel in charge of nondestructive testing operations must have adequate background knowledge for resolution of complex problems in establishing tests and interpreting results.

Background Knowledge. It is important that a NDT supervisor be well versed in all the available NDT methods, their applications and limitations. In addition, the reliability of the methods and their correlation with desired material and product characteristics are very important.

Knowledge of the product design, purpose, and function together with process details may enhance application of test methods by supplying information regarding the importance of the test interpretations and possible sources of discontinuities, faults, and/or defects that could cause product failure.

Familiarity with all policies, local, industry-wide, governmental, or safety and environmental agencies may affect the inspection methods chosen. Although these policies may be seemingly unimportant at times, they can have very strong implications and influences.

Ability To Communicate Is Always Important. The ability to communicate in both written and oral manner cannot be overemphasized in consideration of successful job accomplishment, particularly in supervisory positions.

It is necessary for a NDT supervisor to direct, instruct, and manage the personnel of that group.

It is essential that cooperation be maintained with other manufacturing or working personnel with whom the projects are associated.

There is need for concise accurate reports to higher management. These reports, by the way, particularly when written, are the principal evidence by which reputation and advancement are developed.

In addition, NDT supervisory personnel are often called upon to interface with customer personnel in solving problems and interpreting test results.

QUALIFICATION AND CERTIFICATION OF NDT LEVEL III TESTING PERSONNEL

The American Society for Nondestructive Testing, in its Recommended Practice No. SNT-TC-1A entitled "Personnel Qualification and Certification in Nondestructive Testing" indicated the responsibilities and capabilities of Level III nondestructive test personnel in the following words:*

"An NDT Level III individual should be capable of establishing techniques and procedures; interpreting codes, standards, specifications, and procedures; and designating the particular test methods, techniques, and procedures to be used. The NDT Level III should be responsible for the NDT operations for which qualified and to which assigned, and should be capable of interpreting and evaluating results in terms of existing codes, standards, and specifications. The NDT Level III should have sufficient practical background in applicable materials, fabrication, and product technology to establish techniques and to assist in establishing acceptance criteria where none are otherwise available. The NDT Level III should have general familiarity with other appropriate NDT methods, and should be qualified to train and examine NDT Level I and Level II personnel for certification."

SOURCES OF TECHNICAL INFORMATION AVAILABLE TO NDT PERSONNEL

Regardless of the sources of technical knowledge by which NDT personnel obtain their positions, perpetual updating and extension of that knowledge is essential.

Continual Improvement of Entire Knowledge Base Needed. Additional knowledge of NDT techniques, particularly new developments, is very important. Accompanying this, however, should be broadening of knowledge, as well as keeping up with new developments, in the field where the NDT techniques are applied.

Greater familarity with engineering materials, manufacturing processes, changing designs and service requirements, management policies, government requirements, environmental impacts, personal safety and other areas can be vital to successful application of NDT.

Many Sources of NDT Information Available. Personal ambition and expenditure of time permit information to be gained in a variety of ways.

Activity (not simply attendance) in suitable technical societies can be a good source of new methods and applications of both testing and manufacturing techniques.

---

*The statement of NDT Level III requirements is taken from the June 1980 Edition of ASNT Recommended Practice No. SNT-TC-1A, as illustrative of general requirements. Note that future changes may occur in SNT-TC-1A, and the current document should be referred to in all cases.

Attendance at manufacturers' training courses and short courses offered by universities often supplies good up-to-date information.

Studying current technical literature, including advertising, often provides leads for new applications.

Personal contact with other NDT peronnel is a vital source of information that may lead to applications of entirely new test techniques or to use of well-known techniques in entirely new ways or to new applications. Personal interchange of knowledge may also prevent costly errors and disasters that others have learned to avoid.

This personal relationship can sometimes be established locally in large organizations, but, more often is best accomplished by visits to other organizations for a specific purpose, with contacts made during attendance at local, regional, and national technical society meetings, and by arranged discussions with representatives of equipment manufacturers.

The study and use of standard references such as ASNT *Nondestructive Testing Handbook (2 volumes)* published in 1959 and reprinted in 1963 and 1977, and Volume 11 of the eighth edition of the *Metals Handbook* on Nondestructive Testing and Quality Control, published by the American Society for Metals in 1976, should be valuable to all nondestructive testing personnel.

Other organizations such as the American Welding Society have prepared material on nondestructive testing. A variety of NDT materials is available through the American Society for Testing and Materials. A comprehensive list of material related to nondestructive testing and related subject matter is available through the National Headquarters of the American Society for Nondestructive Testing (ASNT), 3200 Riverside Drive, Columbus, Ohio 43221.

Many Sources Of Materials and Properties Information Available. It would be virtually impossible to name or list all of the important sources of information dealing with materials, their properties, and their processing. Any of the sources could be important to NDT personnel for aid in solving specific problems; aside from that, it is important the NDT personnel, particularly at Level III, have a basic, broad understanding of these subjects.

The purpose in use of NDT is to locate various faults in materials and products that have been processed to at least some degree. The flaws, defects, or imperfections that may be located by inspection may be a result from the original material, caused by the processing used, created by some human error, or be a result of some combination of these. In most cases it is important for the NDT inspector to be able to locate the source of the problem when it exists, or even when a fault is not located, know that the possibility of one exists. An understanding of the materials, the processes, and the possible interactions between them is therefore a "must."

In performing their work, NDT personnel are of necessity in close contact with manufacturing personnel and, to obtain the needed cooperation, must be able to speak a suitable language and understand some of the problems, requirements, policies, and operations involved in the manufacturing process.

NDT is also performed in the field during or after construction and sometimes after failure to determine the real reason for failure. It is important therefore to also have an understanding of the effects of various environmental factors on material properties.

To prevent recommendation of lengthy publications written with great depth of treatment or the use of multiple publications, this text has been developed with the hope that a brief, concise treatment of the materials and processing subject area will provide maximum benefit to those that need a broad background knowledge. When more detailed knowledge is needed, it may be found in many sources in whatever depth needed.

MANUFACTURING: MATERIALS AND PROCESSES

The text material in the following chapters presents an introduction to the many facets of industry which involve needs for nondestructive evaluations of materials, relating to control of properties, effects of processing, problems in welding and assembly, finishing and protection, and serviceability of engineering materials. The response of metallic materials to various stages of manufacture, construction, or service life can vary widely as a function of chemical composition, thermal treatment, mechanical working, surface conditions, presence of discontinuities, and other material characteristics. Most nonmetallic materials have quite different properties and response characteristics, and so their manufacture, applications, and test requirements may differ from those of most metalic materials.

Material Characteristics Often Critical. NDT personnel need familiarity with these many material characteristics in order to evaluate their suitability for service, through nondestructive testing. They also should be prepared to advise management of possible methods for alleviating undesired response characteristics of materials, especially during processing, manufacturing, and assembly operations. Determining the source or cause of defects is frequently necessary in order to eliminate these defects from production parts. Often, the causes of defects lie in early forms or stages of material production or processing. In some cases, these prior processes control the response of materials to later processes, during which defects or failures are induced.

MATERIAL FAILURES

Some products are purely decorative in use or have such low strength requirements that they are inherently over-designed from the strength point of view. These may require inspection to be certain that they will maintain their as-manufactured qualities such as color, polish, stability, etc.

Products more likely to need careful testing and evaluations are those used in load carrying applications where failure may involve loss of use, expensive repair, or danger to other products, structures, and even life. Although the manufactured item is a product, it is the material of that product that may fail, so material failure types and causes are of interest.

Material Failure Definition. The simplest definition of failure is that the item of interest becomes unusable, but there are several ways in which a product may become unusable. It is usually important to know the type of failure that might be expected in order to know for what to inspect, how to inspect, how to eliminate the fault, and how to assess the risk of failure.

Complex units with moveable parts may become inoperable because of failure of some minor element. An automobile, for example, may not run for lack of fuel, a tire goes flat, or the ignition is out of adjustment.

There are two generally accepted types of material failure: one is the easily recognized *fracture* or separation into two or more parts; the second is the less easily recognized *permanent deformation* or change of shape and/or position.

Although complete fracture is unmistakable, an incipient type which will be discussed in connection with "fatigue failure" with suitable inspection methods can be determined before complete failure occurs. Fracture failure in some complex structures may also become *progressive*. An example of progressive failure would be the release of load by some weak component in a structural configuration such as a large bridge. The released load must be absorbed by neighboring structural elements. Unless these neighbors can spread the new load and become stabilized, they will become overloaded and, if stressed above their elastic limit, will deform, crack, or fracture, causing additional load to be passed along in a way that causes the entire structure to collapse almost immediately.

A simple example of progressive failure would be the breaking of a gear tooth in a mechanical power system. The following gear tooth is then subjected to shock (impact) loads which increase the stress levels so that failure is more likely. In this case also it is probable that if power is maintained, all the gear teeth will be broken off in a short period of time.

Material Failure Causes. Products and structures may be subject to a number of service conditions. Imposed loads may be static (stationary or fixed) or dynamic (varying). The use environment may contribute corrosion, vibration, or temperatures and pressures higher or lower than normal. The product may also be subject to abuse. Mechanical failure is always a result of stresses, above some critical value for each material that cause deformation or fracture. Such excessive stresses are set up by some combination of material defect, excess load, improper type load, or design error.

1. As far as failure is concerned, static loads sometimes include dynamic loads that are slowly applied. The principal reasons for failure under static loads include large discontinuities (both internal and external), poor dimensional control during manufacturing, massive overloading during use, and unsatisfactory original design or combination of these factors.

2. Dynamic loads are varying loads that can be single-directional or multi-directional with multi-directional loading being more serious as a cause for failure. When the cycles of loading become high (usually millions but dependent on the material), failure can occur at stress levels far below those determined by static load tests. Although millions of cycles seems high, there are many applications, such as a rotating shaft under bending load, where millions of loading cycles can be reached in a relatively short time.

As pointed out, slow or low frequency dynamic loading is similar to static loading except that even low frequency loads applied suddenly create a condition of shock which can cause failure at a level lower than normally expected.

3. Service at high temperature reduces most of the desirable material properties of metals including the ability to support load. The tendency for *creep* also increases with increased temperature. The temperatures at which property values become critical depend upon the particular material and the previous treatment it has received.

Most metallic materials also exhibit a brittle characteristic (much like cold glass) through a *transition temperature* range usually at lower temperatures.

4. Pressure creating stress above a material's elastic limit may cause material flow (plastic flow), distortion, and cross-sectional weakening, effects that would be intensified at elevated temperatures. Fluctuating pressures of high frequency create dynamic loading that may decrease safe operating levels. It was reported that several early day pressurized cabin aircraft failed by fuselage skin failure due to this cause.

5. Corrosive environments or a combination of materials that cause corrosion can produce failure in two ways. The corrosion may actually reduce the amount of material available to carry load, but even more important in many cases is that the corrosion may create small discontinuities which serve as stress risers that become the nucleus for fatigue failure.

6. Many structures and systems are subjected to vibration during service. Included are transportation equipment, machines, and devices that have moving parts. In addition, some structures may vibrate because they are excited by some outside influence. Stresses from vibration may be superimposed on stresses from other loading sources. The principal problem created by vibration is the introduction of cyclic loading leading toward fatigue failure.

7. Excess loading from abuse may be accidental but nearly always has the human element as a source. Control is attempted by use of design factors of safety,

usually based on yield strength and ranging from slightly more than one to five or more. Factors of safety are applied during design dividing the nominal allowable strength of the material by the safety factor.

8. Use of equipment in improper environment may be considered a type of abuse but is sometimes unavoidable. The main problems experienced are high temperature which may decrease material strength and corrosive conditions which may initiate failure, or, as a minimum, decrease the aesthetic properties of the product.

9. Another form of abuse is improper maintenance, including lack of suitable lubrication of moving parts, and improper cleaning and finishing which may permit corrosion to begin.

10. Some materials deteriorate with age and that deterioration accelerates with relatively small increases of temperature. Many plastics, most glass, and some metals can develop a brittle characteristic with natural aging and become particularly susceptible to failure under shock loading.

Suitable Tests Essential. If failures are to be prevented by use of nondestructive tests, these tests must be selected, applied, and interpreted with care and on the basis of valid knowledge of the failure mechanisms and their causes. The purpose of the nondestructive test design and application should be effective control of materials and products, leading to satisfactory service without premature failures or objectionable damage.

Nondestructive Tests Are Performed on Materials. It is rather obvious that knowledge of materials and their properties should be important to any nondestructive testing person. Most test procedures are designed to allow detection of some kind of interior or exterior fault, or measure some characteristic, of a single material or group of materials. The source of the problem may be a discontinuity, or it could be a material that is chemically incorrect, or that has been treated in such a way that its properties are not suitable.

Discontinuities. The term "discontinuity" is used to describe any local variation in material continuity including change in geometry, holes, cavities, cracks, structure, composition, or properties. Some discontinuities such as drilled holes, or irregular surface shapes, may be intentionally designed and should have been given full consideration by the designer. These normally do not require testing unless the material is being used under critical conditions or trouble has been experienced in service.

Other discontinuities may be inherent in the material because of its chemical make-up and structure. Structure refers to the three-dimensional atomic arrangement in which solid metals and other engineering materials exist. This type of discontinuity can vary widely depending on the particular material, the treatment it has received (intentional or incidental), and its environmental exposure. Because this type discontinuity can vary so much in size, distribution, and intensity, testing to determine its effect may be in order. An exception would be when the discontinuities fall well within the limits expected by the designer and when there is little probability of their being affected by outside influences.

Discontinuities therefore are not always bad or hazardous and may even sometimes be needed in the design or may be helpful in some kinds of processing.

Defects. When any discontinuity, single or multiple, is of such size, shape, type, and location that it creates a substantial chance of material failure in service, it is commonly called a "defect." Finding defects is one of the most frequent objectives of NDT. It must be understood, however, that a fault that is a defect under one set of conditions may be only a simple discontinuity that is not harmful in a different application.

For example, cast iron is a material that is "loaded" with discontinuities consisting of free graphite flakes, voids (both microscopic and macroscopic), and sometimes cracks or tears where the atomic structure is completely separated. Because of this internal structure, the material is never intentionally used under more than relatively small tensile loading (then usually the result of a bending load), but is found to be very satisfactory in many applications where the loads are principally compressive. Cast iron, because of the kind of use it receives, is seldom the objective of NDT, although it possibly contains more internal flaws than any other commonly used material.

Discontinuities May Grow Into Defects. In light of the above statements, it should be pointed out that under some conditions, discontinuities believed to be harmless can change into serious defects that can cause disastrous failure. This is most likely to occur under service conditions and could be because of the effects of fatigue or corrosion, especially when accompanied by cyclic loading. A small discontinuity started by corrosion, a slight scratch, or a discontinuity that is inherent in the material, may develop into a crack from the stress concentration that, under varying loads, propogates with time until there is no longer sufficient solid material to carry the load. Sudden total failure by fracture then occurs.

An example of this type failure is the collapse of the Silver Bridge across the Ohio River at Point Pleasant, West Virginia in 1967. Many aircraft parts require careful nondestructive testing and evaluation because they are designed with high stress levels and low safety factors to keep down weight; very small discontinuities may develop into failure defects.

Processing Affects Materials Properties. To this point, it would appear that all the emphasis would be on materials, their structure, and their properties. Regarding NDT this is true, except that it must be remembered that the processing of those materials from the raw state through to the completed product has a large influence on the characteristics of the final material.

Some processes such as heat treating are for the expressed purpose of affecting material properties. Other processes such as casting, welding, forming, and machining makes use of heat and/or deformation forces to perform their function and the reaction of the material is the same as though the procedure were carried on to change the material properties.

To some degree, knowledge of the processing is therefore necessary in order to understand the effect on the material, to evaluate the material properties, and to trace down the source of problems.

PURPOSE FOR USE OF NDT

A critical task for persons responsible for nondestructive testing operations is often the determination of the true reasons why nondestructive tests have been requested, specified, or needed. The interpretation of test indications depends critically upon the purpose of the tests, and this often determines the stage of manufacturing or assembly at which tests should be done.

Ultimate Purpose—Reliability, Serviceability. In the preceding paragraphs, the common assumption has been made that the tests are to be used to assure reliability and to prevent premature failures of materials, parts, or assemblies during their intended service. In general, this serviceability is the ultimate purpose of most nondestructive tests. However, it is often the case that the test itself does little to predict the serviceability or safety of the final product, assembly, structure, or system. In this case, there is no way to determine that a discontinuity or material condition constitutes a dangerous "defect" which may lead to premature failure in service. This by no means invalidates or countermands the need for the nondestructive test. Other valid reasons for nondestructive testing may include the following.

Specific Purposes for NDT.
1. Identification or sorting of material.
2. Identification of material properties and the reliability associated with their existence.
3. Indication of proper material and suitable quality control during processing in order to prevent further costly processing.
4. Tests to assure completeness, proper dimensions and geometry, and proper relationships among assembled components.
5. Tests during service to discover initiation of possible failure before it actually occurs.
6. Diagnostic tests after failure to determine the failure reason. This knowledge might be useful for product design change, test method change, quality control records, and for records to combat possible product liability suits.

NDT IN FRACTURE CONTROL

Fracture Control Design Philosophy. When using factors of safety in design, it is assumed that competent judgement in providing a safety factor can provide high assurance of the safe life of a product or structure. At the same time, consideration must be given to conservation of material. Historically, many complex structures and machines have survived admirably under such design concepts. However, others have failed from seemingly inexplicable causes, some with costly and disastrous consequences.

Most design procedures still are based upon over two and a half centuries of using Hooke's law which relates stress and strain in elastic bodies. The safety factor is applied to a value of strength that the material used in a design is presumed to possess. The value of strength is that which is hoped to be representative of the material used under the assumption that the material is continuous and has uniform properties throughout. In this design process, discontinuities can be accounted for, *if it is known they exist,* by their effect on reduction of cross-sectional area available to sustain the applied loads, or their effect on the local volumetric strength of the material.

Conventional design practices incorporating safety factors ordinarily will succeed if:
1. the material used will not be accidentally or otherwise overloaded,
2. the variations in ordinary loads will be as intended,
3. the environmental factors are properly anticipated,
4. unknown or undetected discontinuities will not grow to a critical size.

Notable catastrophic failures have been analyzed and invariably found to involve an oversight in one or a combination of the above factors that caused failure. Sudden, complete, and unsuspected fracture failures occur in components and structures when a crack or other defect reaches some critical size and rapidly propogates.

Where the consequences of unexpected failure are unusually expensive in terms of public safety and/or money, more and more designs of such critical nature require consideration of fracture-resistant qualities in materials selection and usage. Implicit in frature control design criteria is the need, by some means, to assure that unexpected flaws of some specific critical size are not present when the product is introduced into service. Nondestructive testing and proof testing are the principal means upon which such assurance is based. Proof testing always involves some risk that the test itself will either cause immediate failure during the test, or cause flaw growth to a point below actual failure but beyond the point where additional flaw growth can be safely sustained during the service life of the component in question.

Nondestructive testing is heavily relied upon as the basis of assurance against the presence of flaws large enough to either cause immediate fracture or to grow large enough to cause later premature fracture. Often, the successful implementation of a fracture control design depends solely upon the reliability of nondestructive testing to detect flaws in otherwise sound components so that repairs or other corrective measures can be taken.

Fracture Mechanics. The study and description of a material property related to fracture resistance is fracture mechanics. Fracture mechanics analytical studies and fracture mechanics testing attempt to quantify the fracture toughness of a material. The property of fracture toughness is expressed quantitatively in terms of the stress intensity factor, K. K is a measure of the intensity of the stress field surrounding the tip of an ideal sharp crack in a linear elastic material. The stress intensity increases rapidly in the small, localized volume surrounding the crack tip when the crack faces are pulled apart by loads imposed normal to the crack plane. The local stress at the crack tip is greater than the stress in unflawed areas. If a critical value of this stress is exceeded due to increased load, the crack can advance by stress-rupturing the material just ahead of the crack tip. In brittle materials, this sudden, small stress-rupture may release enough energy into the new crack tip region to permit the process to continue in a rapid, uncontrolled manner. At this point, structural stability is compromised and the part fails completely and suddenly, often catastrophically.

The most common condition under which K, the stress intensity factor, is considered is that of plane strain. Plane strain is the condition where strain surrounding the crack tip is zero in the through-thickness direction. This condition results when the material is relatively thick compared with crack size and the contained crack is under severe tensile constraint. Under such conditions, the property of interest is the critical plane stress intensity factor, K_{Ic}. Knowing the value of K_{Ic} for a particular material under particular conditions enables the designer to calculate the critical flaw size. At the design stress, a flaw larger than this size can trigger brittle fracture.

Prior to the time that serious consideration was given to fracture mechanics, some materials selected for their high strength capability failed under relatively low loads. Flaws initially small in size were found as the origins of failure, having propagated to critical size. In some instances, the initial flaws wre smaller than could have been reliably detected by nondestructive testing.

Inspection Reliability. Fracture mechanics presumes the presence of flaws in finished structural elements. Furthermore, through analysis and testing, fracture mechanics predicts the size of flaw which can cause brittle failure either as an initial critically-sized flaw or as a smaller flaw that can grow to critical size under cyclic loads. In order to realize an advantage from such an analysis, some form of inspection or testing must be applied to the product. Proof testing and nondestructive testing are the most adaptable means to detect flaws. There are advantages and disadvantages of each form of testing.

The risks in proof testing have been previously mentioned; in addition, proof testing often is quite expensive. Nondestructive testing can also be expensive and, without special care, is not routinely called upon to provide assurance that flaws exceeding a certain size are not present in a test object. In most cases, nondestructive testing is applied as either a qualitative or semiqualitative tool in inspection and process control. In fact, most nondestructive tests provide only indirect indications of actual discontinuities. Nondestructive testing personnel are truly challenged to answer the questions posed by engineers using fracture mechanics concepts:

1. Will the inspection procedure to be used *guarantee* that *all* flaws greater than some critical size will be detected?

2. What is the largest flaw that can escape detection using a particular inspection procedure?

With the present state-of-art of NDT and for some indefinite future time, these questions cannot be answered with the precision desired by design engineers wishing to use fracture mechanics concepts. However, an approach presently accepted uses statistical methodology to define inspection reliability.

Probability of Detection/Confidence Level. Given an inexhaustible continuum of flaw sizes and shapes for all the materials of interest along with substantial financial resources, each inspection procedure could be tested with flawed specimens. Straightforward statistical methods could be used to demonstrate the probability that a particular procedure would (or would not) detect flaws of given sizes. In simple terms, if a large number of specimens each contained a flaw of the same size and all were subjected to a given inspection procedure, the ratio of flaws detected to total number of flaws looked for would constitute the probability of detection. For example, if 100 flaws of the same size were present and 90 were detected, for the circumstances of this particular experiment it could be stated that the probability of detection of flaw size X in a given material is 0.9 or 90%.

Based on this example, repeated experiments on additional flawed specimens could be conducted. Instead of 100 flawed specimens being examined, 1000 or even 10,000 could be tested. If 9,000 out of 10,000 flaws were detected, the confidence would be increased that the true probability of detection is 90%.

In practical situations, 100 flawed specimens would be a luxurious sample, indeed. It is possible, however, to estimate from a limited sample size the probability of detection for a larger population. In the original example, where only 100 units were tested, it would be useful to know how precisely true is the inference that the detection probability is 90%. In statistical analyses, the term "level of confidence" refers to the probability that the 90% detection probability inference is truly valid.

The interpretation of a statement that a particular experiment produced a 90% probability of detection with 95% confidence is that there is a 5% probability that the 90% probability of detection is overstated.

NDT Demonstration Programs. Pioneering efforts to include meaningful fracture control criteria into structural design first appeared in the specifications for military aircraft in the late 1960s. Since then, fracture control criteria have been applied to aircraft, spacecraft, nuclear components, pipelines, and pressure vessels. The most ambitious programs to date have included several military aircraft and NASA's Space Shuttle. Typically, the overall vehicle specifications have required the contractors involved to demonstrate inspection capabilities to detect certain size flaws in components designated fracture critical at 90% probability of detection, 95% confidence level.

Several approaches have been used by the contractors to provide the required demonstration. Generally, some economically practical numbers of fatigue cracked specimens are prepared and intermingled with unflawed specimens. Well-defined procedures are used by inspection personnel under production conditions. The parts used in the demonstration are usually entered into the normal sequence of inspection so that the inspection personnel are not biased toward particular awareness that a demonstration is being conducted. In other cases, only certain inspection personnel have been classified as being qualified to inspect fracture critical components. Such components are clearly marked and channeled specifically to their selected inspectors. In any event, through experiment design and statistical analysis, the outcome of NDT demonstration programs is intended to provide a workable interface between the needs of fracture mechanics and the realities of the applications and limitations of NDT.

Introduction to Manufacturing Technology 2

Webster defines "manufacture" as "to make by hand, by machinery, or by other agency; to produce by labor, especially now with division of labor and usually by machinery."

Such a definition is all-inclusive. It covers the making of foods, drugs, textiles, chemicals, and, in fact, everything made usable or more usable by the conversion of shape, form, or properties of natural materials.

Special interests have developed in the mechanical and industrial phases of industry concerned with the making of durable goods of metals and plastics. The majority of metals and some other materials fall in a class that is often referred to as *engineering materials*. Characteristic of this group are the properties of relatively high hardness, strength, toughness, and durability. Glass, ceramics, wood, concrete, and textiles, although they may compete with metals in many applications, have usually been excluded from these structural materials because of a difference in the combination of properties, a difference in processing requirements, and a difference in type of goods produced. The list of so-called engineering materials continues to grow with the addition of new metallic combinations, plastics, and even materials that have been previously excluded from the list, as they are developed with better properties or used in new applications.

Present interpretation of the term *engineering materials* includes most metals and those plastics that are solids and have reasonable strength at room temperature. This book will be concerned with these materials and the processes that are used to shape them or change their properties to a more usable form.

HISTORY

The growth of industry in the United States is typical of industrial development throughout the world. Early settlers were concerned primarily with food and shelter. Most manufactured goods were imported but some manufacturing was done in the family units. Eventually, as conditions were stabilized, efficiency improved and excess goods were available for sale and trade. The factory form of industry finally resulted, under control of single families. Some of these still exist but most have changed to corporate enterprises under ownership of many individuals.

Early Manufacturing. The first manufacturing was devoted mainly to agricultural and military needs. One of the earliest industrial operations to grow to large size was the reduction of ore to metal. By its very nature, particularly for ferrous metals, this process is not adaptable to very small operations. The trend in this industry to increasing size has continued to the present. A few very large corporations produce nearly all of the basic metals, even though there are many small fabricators.

Interchangeability. The Civil War and the expanding frontier created much incentive for the manufacture of firearms. Many will remember that the first example of true interchangeability and the development of better transportation following the Civil War resulted in rapid growth of production goods. Many of the products were considered luxuries at the time but since have become necessities to the modern life style.

Importance of Manufacturing. Manufactured products are an integral part of everyone's life, but most persons do not realize the great amount of investment and labor that makes those products possible. Realization comes with thought that almost every activity, regardless of field, is in some way dependent on hardware produced by the manufacturing industry. Approximately 25% of the gross national income is spent for manufactured goods and about the same proportion of the United States' working force is employed in the manufacturing industry.

INDUSTRIAL RELATIONSHIPS

COMPETITION IN INDUSTRY

In the American way of life, the profit motive is the root of most business, including manufacturing. The system presumes direct competition, so that if a number of companies are engaged in the manufacture of similar products, the sales volume will be in proportion to the product quality, promotional activities, service policies, and price. The cost of manufacturing therefore becomes of prime importance, for the company that can produce at the lowest cost and maintain quality can spend more for sales activities, can sell at a lower cost, or can make a larger profit per sale than competitors in a less fortunate position. For this reason industry is continually engaged in a battle to lower production costs and to gain this favored position.

Direct Competition Limited. Because of the complexity of the overall manufacturing operation, many decisions are, of necessity, rather arbitrary. For nearly all products, there are many alternatives of design, materials, and processing that will satisfy the function the product is to have. For many products, direct sales-price comparisons are not adequate, for different demands for similar products made of different materials or having different designs may exist. The purchaser is truly the final decision-maker, which makes advertising and sales promotion a most important phase of the business.

Adequate time is often not available to study the effect of a design on the market or to investigate all the possible processes of manufacture, particularly for new products. Sometimes, to determine the exact material that would serve best even for a fixed design is too time consuming. In any case, reasonable decisions must be made, and when absolute knowledge is not available, they are based on past experiences of similar nature. Because of the interrelationships existing in manufacturing, accurate decisions will depend not only on exact knowledge of a specific area but also on knowledge of interaction from related areas.

PERSONNEL

Several kinds of workers are needed in any manufacturing operation. Some work directly with the product, and some are only indirectly connected with the product but are more concerned with the organization producing the goods. Those directly connected with the product include the designer, those responsible for choosing the processes, establishing control over the operation, and supervising the manufacturing, and the machine and equipment operators who perform the actual work of converting raw material into useful objects. Each of these, to function effectively in his job, must have varying degrees of knowledge concerning the product requirements, the material properties, and the equipment limitations. Most jobs directly connected with the product call for specific knowledge in depth concerning certain phases of the work and more general knowledge of related areas.

Products, from the simplest single part items to the most complex assemblies costing millions of dollars each, go through a series of chosen steps of manufacture as they proceed from raw material to completed useful products. In order to conserve energy, material, time, effort, and to reduce cost, it is necessary at each stage of product development that qualified personnel examine the processed material to insure that the final product has the quality and reliability expected from the design. A large part of the manufacturing effort therefore is in addition to modifying material and adding to the product development. Essentially all products require a degree of inspection of the material to see that it conforms to the requirements that provide a high quality product.

Although not normally classed as direct labor, sales personnel usually must have complete familiarity with the product and its manufacture. They are called upon to recommend, compare, troubleshoot, and even install a product.

Indirect. Other personnel are only indirectly connected with the product or the manufacturing operation. These include most workers in administration, accounting, finance, purchasing, custodial service, and other support areas. The personnel who work in these areas may be highly skilled or trained in their own field. They do not need extensive technical knowledge of the product or its manufacture. However, they may still make decisions that are far-reaching in effect on the products. Therefore, they do need broad understanding of the product and the manufacturing facility.

NOMENCLATURE

The ability of personnel from one area of manufacturing to discuss and understand problems with people from another area will depend directly on their knowledge of the nomenclature used in the area of concern. A designer, to discuss intelligently with a production man the effects of various design changes on the method and cost of production, must be able to understand and use the language of the production man. In most cases, he needs to know at least the names of the various machines and tools that might be used and have some understanding of their capabilities. In the final analysis, the problems of the production of a product become the problems of the machine and equipment operators. The loyalty, cooperation, and respect for supervision of these operators, necessary for the proper solution of production problems, can be gained only when a full understanding exists between the two groups. Of necessity, this understanding must be based on suitable language, including proper terminology, even to the point of using local terms and nicknames when **appropriate. Similarly, NDT personnel must communicate with production and other personnel.**

SUBJECT MATTER

Even with the limitations that have been placed on the term *manufacturing processes* for use in this text, many possible variations of content and organization of subject matter exist. The principal objective of this text will be to present a broad discussion of the materials used in manufacturing and the principal processes by which these materials are made into usable products. The subject of *materials and manufacturing processes* is truly a single subject when the orientation of discussion is toward the end product that must be manufactured to fulfill some function. Although the attempt has been made in this book to show this singleness of subject matter, it is still necessary to treat specific areas as specific topics. Similarly, manufacturing plants are normally divided into areas in which the equipment and personnel concentrate on particular manufacturing operations. For example, a foundry may produce only iron castings of a certain weight range because of specialized experience and equipment.

MATERIALS

An understanding of materials is important to any manufacturing procedure. One or more materials are required for any product, and most can be processed in a number of different ways. However, for many materials, the processing possibilities are very limited, and the process may be dictated by the particular material chosen.

Properties. The practical differences between various materials is in their *properties* or combinations of properties. Compared to many other materials, steel is hard and strong and may be chosen as a manufacturing material for these reasons. Steel is elastic to some extent. However, if elasticity is the important property of interest, it may be necessary to choose a material like rubber for the application. An intelligent comparison of materials depends on precise meanings of the terms used and an understanding of how properties are defined and measured. Some properties are defined by tests, such that the results may be used directly as design data. For example, from a standard tensile test, the modulus of elasticity of a material may be determined, and a designer can use this value to predict accurately the deflection of a certain-size beam under known loads. On the other hand, many properties are defined no less specifically but in a more arbitrary manner, which makes the use of the test results for calculation difficult or impossible. However, the tests still provide the opportunity for accurate comparisons with data obtained from similar tests from other materials. For example, hardness measurements may give an indication of relative wear resistance for different materials, or hardness numbers may correlate with tensile strength for a given material, but the number values can seldom be used directly in computation for design loads.

Property Variations. Each elemental material has at least some properties different from those of all other elemental materials. Some or all of the properties of an element may be changed by the addition of even small parts of another element. In many cases the properties obtained from the combination will be better than those of either element alone. In a similar manner, the properties of elements or combinations can be varied by the type of treatment given the material. The treatments that affect properties are often intentionally selected for this purpose. However, the properties are no less affected, often in an undesirable way, by the processes being used with the objective of shaping the material. Sufficient knowledge of the relationship between the properties and the processing of materials may permit the improvement of the properties as a natural result of the processing for a different main objective. Reducing the cross-sectional size during the shaping of most metals results in an increase in hardness and strength that may be undesirable if the metal must undergo further deformation processing. In many cases, this increase in hardness and strength that occurs as a result of the processing can be beneficial and part of the product design.

PROCESSES

Manufacturing consists of converting some raw material, which may be in rough, unrefined shape, into a usable product. The selection of the material and the processes to be used seldom can be separated. Although in a few cases some unusual property requirements dictate a specific material, generally a wide choice exists in the combination of material and processing that will satisfy the product requirements. The choice usually becomes one of economic comparison. In any case, a material is usually selected first, sometimes rather arbitrarily, and a process must then be chosen. Processing consists of one or many separate steps producing changes in shape or properties, or both.

Shape Changes. Shape changing of most materials can be accomplished with the material in one of several different forms or states: liquid, solid, or plastic. Melting of a material and control of its shape while it solidifies is referred to as casting. Reshaping of the material in the plastic or semisolid form is called molding, forging, pressworking, rolling, or extrusion. Shaping by metal removal or separation in the solid state is commonly performed to produce product shapes. If the removed material is in chip form, the process is machining. The joining of solid parts by welding usually involves small localized liquid areas that are allowed to solidify to produce a complete union between solid parts.

Energy Form. The material condition and the energy form used to effect these shape changes may vary. As noted, the material may be in a liquid, solid, or plastic form. The energy may be supplied in the form of heat, mechanical power, chemical reaction, electrical energy, or, as in one of the newest procedures, light. In nearly every instance, one principal objective is shape changes, but usually part of the energy is consumed in property changes, particularly in those processes involving state changes or solid deformation. Different materials react differently to the same energy system, and the same materials react differently to different energy systems.

Process Effect on Properties. Many concepts and fundamentals in reference to materials are common to different kinds of processes. When studied in connection with the material, these concepts, then, can be applied regardless of the kind of process by which the material is treated. The metallurgical changes that take place during solidification during casting are of the same nature as those that take place in fusion welding.

Auxiliary Steps. The completion of a product for final use generally includes the various finishing procedures apart from basic shape-changing processes. The dimensions and properties that are produced by any process are subject to variation, and, in practically all cases, some *inspection* of nondestructive type is necessary for controlling the process and for assuring that the final product meets certain specifications as to size and other properties. As one of the final steps, or sometimes as an intermediate step, control of properties by *heat treatment* or other means may be necessary. The final steps may also require surface changes for appearance, wear properties, corrosion protection, or other uses. These steps may involve only the base material or may require the addition of paints, platings, or other coatings.

Few finished products are constructed of single pieces of material because of the impracticality of producing them at a reasonable cost. Also, it is frequently necessary that properties that can be obtained only from different materials be combined into a single unit. The result is that most manufactured articles consist of *assemblies* of a number of separate parts. The joining of these parts can be accomplished in many ways, with the best method being dependent on all the factors of shape, size, and material properties involved in the particular design.

ECONOMICS

The private ownership systems of business and industry in the United States are profit motivated. In a competitive market, the manufacturer who makes the most profit will be the one who has the best combination of design, materials choice, and manufacturing processes. Ultimately, most decisions become a compromise between the most desirable from a design, life, and function standpoint and the most practical from a production and cost standpoint.

Design. The designer must not only know the functional requirements of the product but also have some knowledge of the probable market demands for various levels of quality and appearance. He certainly must be familiar with the mechanical properties of the various materials he might choose.

Less obvious at times is the importance of the part the designer plays in the selection of manufacturing processes. If the designer designates a sheet-metal housing for a radio, obviously, the housing cannot be a plastic molded part or a die casting. If he specifies certain tolerances, these not only may dictate that a certain dimension be achieved by machining but also may even dictate the specific type of machine to be used. Clearly then, in every case, the designer's choices of materials, shapes, finishes, tolerances, and other factors restrict the possible choices to be made in the manufacturing process. **The designer may also specify the NDT criteria, thus influencing the choice of NDT.**

Choice of Materials. Engineering materials, metals and others, have properties that vary over wide ranges with many overlaps. Costs also vary widely, but the cheapest material suitable for the product does not necessarily insure the product will have the lowest cost. For example, a lower cost steel substituted for another may satisfy the functional requirements of **the product but may lead to increased inspection costs, thus decreasing or eliminating the margin of necessary profit.**

Quantity. The number of a product that is made can have more influence on the cost than the design or the type of material used. Most manufacturing processes involve both a get-ready, or *setup* cost, and a production cost. The setup cost can range from nothing to many thousands of dollars, depending on the type of process and the amount of special tooling needed. The actual production time for each product is usually inversely related to the setup cost.

Quality. Quality costs money. Higher quality implies longer life, better finishes, better materials, quieter operation, and more precision. These factors all involve greater costs that may be justified by market demand. If not justified, competition will satisfy the demand with lower quality at lower cost.

Inspection. Inspection also costs money to perform, but, in another sense, like advertising, it pays; in fact, it is essential to assure better quality product output and to improve customer relations.

Modern technology has produced much inspection equipment needed for nondestructive testing. However, proper application of inspection methods and interpretation of their test indications is not possible without relying upon qualified nondestructive test personnel. Capable individuals are needed to provide input to the decision processes regarding the integrity and serviceability of the test objects, stemming from the indirect indications provided by nondestructive tests. Such persons must have an adequate background of knowledge concerning the materials and manufacturing technologies involved in their specific industries, and the service conditions to which their products will be subjected, in order to make valid decisions.

ORDER

The enormous quantity of knowledge available about manufacturing processes can be discussed in varying degrees of depth and coverage. The following chapters of this book have been chosen with the hope that the order will seem logical and conducive to maximum learning. The discussion does not go into great detail in the belief that for the purpose of this book broad knowledge of the overall manufacturing system is more important than the development of depth in any special but restricted area.

Materials. As has been indicated in this chapter, the properties of materials are very important and cannot be divorced from the manufacturing processes. The first topic of discussion will therefore be properties, with their definitions, which generally consist of a description of the test procedure used to measure the property, followed by the fundamentals of metallurgy as they apply to the commonly used manufacturing materials and processes. The properties of specific materials will be discussed only as they affect the process choice and as the process affects them.

Processes. The major processes of casting, deformation shaping, welding, machining, and finishing **will be discussed with an emphasis in length and depth commensurate with their use and importance to NDT personnel.** The experiences of many individuals frequently leads to a belief that one area of manufacturing is more important than others, but the interrelationships are such that no one area can exist alone, and the importance of any process in an individual case is entirely dependent upon its relation to the product with which it is associated.

Properties of Materials 3

INTRODUCTION

Because manufactured items are made from materials with various properties, responsible NDT personnel must be generally familiar with engineering materials and their capabilities and limitations. Selection of an engineering material to implement the design of a usable part or assembly requires knowledge of the material's chemical, physical, and mechanical properties. Most structural materials are loaded by external forces which generate high levels of internal mechanical stress within the materials. The reaction of the component to a new stress distribution caused by the development of discontinuities may be critical to its continued functioning. In order to perform meaningful inspections, the responsible nondestructive testing personnel must be cognizant of both the normal material properties and of the effects of discontinuities upon the material serviceability in its intended applications.

As indicated in Chapter 2, the qualities of materials that are of practical interest to manufacturing are measured quantitites called *properties*, as distinguished from the physical makeup of materials called atomic *structure*. Science in recent years has made great strides in determining the atomic structure of materials. Figure 3-1 shows that an atom of iron contains twenty-six electrons and an atom of aluminum contains thirteen electrons, arranged in definite order. The number and the arrangement of particles in each atom actually determine all the properties of any material, and it should be theoretically possible to predict the properties of a material from the structure of its atoms. Physicists and chemists can make some predictions of properties, particularly chemical and electrical, based on structure, but the mechanical properties of greatest interest to a study of manufacturing processes must still be defined and measured by empirical test for each material.

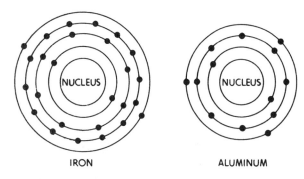

Figure 3-1
Atomic structure

CLASSES OF PROPERTIES

The application to which a material is put determines which of its properties are most important.

Chemical Properties. The chemical properties (reaction with other materials) are of interest for all material mainly because of the almost universal need for resistance to corrosion. Although aluminum is chemically more active than iron, in most atmospheres the corrosion byproducts of aluminum form a denser coating, which acts as a shield to further corrosion, than do the corrosion byproducts of iron.

While the atomic and crystalline structure of all metals gives them high electrical and thermal conductivity compared to nonmetals, individual metals still differ considerably. Aluminum is among the best electrical conductors, while iron, although much more conductive than nonmetals, is a poor conductor compared to aluminum. On the other hand, the magnetic properties of iron make it much more desirable for some electrical uses than aluminum.

Physical Properties. Physical properties for each material are constants associated with the atomic structure. These properties include density (weight per unit volume), crystalline type, atomic spacing, specific heat, cohesive strength (theoretical), and melting point. Iron has a much higher melting point and density than aluminum. Iron is allotropic, meaning it can exist in several different crystalline structures as opposed to aluminum, which always exists in single crystalline pattern. This difference makes possible, for iron-based alloys, methods of property control by heat treatment that are not possible for aluminum. Some aluminum-based alloys may be heat treated for property control, but the reaction is entirely different.

Mechanical Properties. Of most interest to manufacturing are the mechanical properties of hardness, strength, and others that are of prime importance in design considerations for determining sizes and shapes necessary for carrying loads. These qualities will also determine the work loads for any deformation type of manufacturing process. Neither iron nor aluminum in the pure state has many applications in manufacturing because their strengths are low, but their alloys, particularly iron alloys, are the most commonly used of all metals. Both of these materials can be strengthened over their weakest forms by factors of almost ten by suitable alloying and treatment, with alloys of iron being approximately five times as strong as those of aluminum on a volume basis.

Processing Properties. As pointed out at the beginning of the chapter, the properties that have been discussed are actually dependent on the atomic structure of a material, but in practice these properties must be separately measured. In a similar way, different properties that are related to hardness, strength, ductility, and other physical and mechanical properties and that are frequently of even greater importance to manufacturing must in practice be defined by separate tests. These include tests for *castability*, *weldability*, *machinability*, and *bending* that describe the ability of the material to be processed in definite ways. Tests of this type may be developed at any time there is need for determining the ability of the material to meet critical needs of processing, and they are usually performed under conditions very similar to those under which the process is performed.

SIGNIFICANCE OF PROPERTIES TO DESIGN

A designer is necessarily interested in properties because he must know material strengths before he can calculate sizes and shapes required to carry loads, chemical properties to meet corrosive conditions, and other properties to satisfy other functional requirements. Knowledge of processing properties is likely to be of more importance to manufacturing personnel than to the designer, although even he must be able to choose material that can be manufactured in a reasonably economical manner. Many manufacturing problems arise from choice of materials based only on functional requirements without considering which is the most suitable for the processing required. Similar results can occur when inspectability has not been given proper consideration in design.

Material Choice a Compromise. Most products can be manufactured from a number of different possible materials that will satisfy the functional requirements. However, some are more desirable from the product standpoint than others, and one particular material may have the best possible combination of properties. Likewise, all materials can be manufactured by some means, although costs of manufacturing will vary, and there will likely be one single material from which a usable product could be manufactured at lowest cost. Seldom can a material be chosen that has optimum properties for both the product and the manufacturing, so the majority of material choices turn out to be compromises. The final choice may be a result of trial and error tests among several possible best materials and processes. New choices may be required with changes of design, material availability, processes, or market demand.

LOADING SYSTEMS AND MATERIAL FAILURE
LOADING SYSTEMS

Physical loading of material is a result of applying force under one or more simple, basic loading systems. In nearly all cases, even when a piece is loaded by only a single set of outside forces, the internal loads developed are more complex than those applied. However, in many testing procedures this complexity is disregarded, and the forces are treated as though they are uniform thoughout the material.

Stresses. Internal forces, acting upon imaginary planes cutting the body being loaded, are called stresses. For purposes of ease in understanding and comparison, stresses are usually reduced to unit stress by assuming that the force acts uniformly over the cross-sectional area under consideration. The load-per-unit area can then be calculated by dividing the total load or force by the area on which it acts. The common units used for measurement and description in the United States are pounds for force and square inches for area, so unit stress becomes pounds per square inch (psi) or when dealing with large figures, thousands of pounds per square inch (kips/in^2).

With the changeover to the international metric system of measurement, the units for stress become newtons per square meter (pascals) in which a newton is equal to approximately 0.2248 pounds of force (poundals). Conversion of kips/in^2 involves multiplication by the factor 6.894757 to obtain megapascals (MPa). Approximate conversion can be accomplished by use of the multiplying factor 7.

Normal Stresses. Figure 3-2 represents a bar subjected to a pulling force of P. If the load P is uniformly distributed over the ends of the bar, it can be assumed that the internal loads are uniformly distributed. Examination at any plane $x-x$ perpendicular to the line of applied force will show that the crystals along one side of the plane are trying to separate from the adjacent crystals along the plane. This internal force tending to separate the material is known as stress. If the surface area cut by the imaginary plane $x-x$ is A, then the unit stress (s) is P/A, or

Figure 3-2
Simple loading

written as a formula, $s = P/A$. Because in this case the applied force is a pulling force or tensile force, the internal loads are tensile stresses (S_t), and the formula may be written $S_t = P/A$.

Reversal of the external load P would cause the internal stress to be compressive instead of tensile. The unit stress on any plane $x - x$ perpendicular to the line of force would then be calculated from the formula $S_c = P/A$.

Shear Stresses. Tension and compression forces and their resulting stresses are always considered to act normally, or perpendicular, to a plane. A third term, *shear stress*, is used to describe the effect of forces that act along, or parallel to, a plane. No provision has been made for describing forces meeting a plane at an angle. Because, however, an infinite number of planes may be of interest, it becomes necessary to resolve the stresses to various angles to determine critical values and positions. Figure 3-3 illustrates a bar, similar to that of Figure 3-2, with tensile load being applied to the end. As already illustrated, tensile stresses, and tensile stresses only, are set up on any imaginary plane $x - x$ perpendicular to the line of force. If, however, a plane not perpendicular to the line of force is examined, it can be seen that a different situation exists. The imaginary plane $z-z$ is at any angle ϕ. The area cut by the imaginary plane $z-z$ is equal to the area of the plane $x-x$ multiplied by the secant of the angle ϕ. Therefore, the unit shear stress is

$$S_s = \frac{P \sin \phi}{A \sec \phi} = \frac{P}{A} \sin \phi \cos \phi = \frac{P}{2A} \sin 2\phi$$

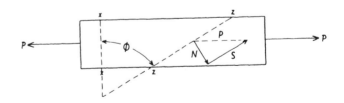

Figure 3-3
Resolved loading

Substitution of the values for ϕ in this formula shows that for zero or 90°, the shear stress is equal to zero. The maximum shear stress occurs when ϕ is 45° and sine 2ϕ is 1, in which case the shear stress, S_s, equals $P/2A$. The maximum value is one-half the tensile stress, S_t, established on a plane that is perpendicular to the applied force.

If, in the preceding case, the external load were compression instead of tension, shear stress would have been developed to the same magnitude and in the same way but opposite in direction and combined with compressive stress instead of tensile stress. Shear stress exists alone only in a bar subjected to pure torsion, that is, a bar being twisted with no tension, compression, or bending present. Shear stresses are important to our manufacturing processes because these are the forces that cause material to shift in plastic flow and permit shape changing by deformation processes.

Bending. Bending loads create a combination of stresses. The concave side of a bent body will be in compression and the convex side in tension with transverse shear occurring along the axis between them. The maximum unit stress will be in the outer fibers of the bent body and is represented by the formula $S_b = Mc/I$ where M equals bending moment, c equals distance from neutral axis, and I equals moment of inertia of the body.

Effects of Stresses. The principal point to be made in this discussion of forces and stresses is that structural designs must be of suitable size and shape and must be made of material with proper strength values to withstand the loads imposed upon them. When a structural member (almost any object) is physically loaded by weight, by pressure from mechanical, hydraulic, or pneumatic sources, by thermal expansion or contraction, or by other means, internal stresses are set up in the member. The size, direction, and kind of stresses are dependent upon the loading system. The magnitude of the *unit* stresses will be dependent not only upon the applied force but also upon the area of material resisting the stresses. As loads are increased, unit stresses will increase to the point where, in some direction, one or more reach critical values in relation to the material. Failure by plastic flow or by fracture can then be expected, depending upon which critical values are reached first. In nearly all cases of fracture failure, the separation of material is preceded by at least a small amount of plastic flow. In those cases in which plastic flow occurs to a large degree, fracture failure will finally result.

TESTING

Testing of material is essential to gain practical knowledge of how materials react under various situations. The ultimate goal of any test is to enable the making of decisions that provide the best economic results. In practice, two general methods of testing are used.

Direct Testing. The only test that supplies absolute information about a workpiece or a material is a test of the particular property of interest conducted on that part itself. In this method of *direct testing*, an attempt is made to use the materials under the exact conditions of practical use, and the test may be concerned with a product, a process, or both. Direct testing is usually time-consuming, and, for the results to have statistical significance, often requires compilation of data from many test samples. The procedure is necessary, however, for those cases in which simpler methods are not available and in which sufficient historical information has not been accumulated to permit correlation between the attribute about which information is desired and some other measurable factor.

Indirect Testing. *Indirect testing* involves the use of such a correlation, such that accurate knowledge of the relationship between the two factors must exist. The ability of grinding wheels to resist the centrifugal forces imposed in use is directly tested by rotating them at higher speeds than those of actual use. Such a test indicates that the wheel strength is sufficient for normal use with some safety margin. An indirect test that is sometimes used for the same purpose can be performed by rapping a suspended wheel to cause mechanical vibrations in the sonic range. A clear tone indicates no cracks. A danger of indirect testing is that the conclusions depend on the assumption that the correlation between the measured factor and the critical factor exists under all conditions. The rapping test for grinding wheels does not give any real indication of strength, unless knowledge of the wheel's history permits the assumption that with no cracks it has sufficient strength for use.

Destructive Testing. A large number of direct tests are destructive. These also are dangerous because the assumption must be made that those materials not tested are like the ones for which test information has been obtained. A portion of weld bead may be examined for quality by sectioning it to look for voids, inclusions, penetration, bond, and metallurgical structure by visual examination. By this operation, this portion of the bead has been destroyed; regardless of the quality that was found, the only knowledge acquired about the remaining portion of the weld comes from an assumption that it is similar to that examined because it was made under the same conditions.

Nondestructive Testing. In addition to the nondestructive feature, these tests almost entirely are indirect tests that require first, correlation with the defects that are being sought, and second, expert evaluation or interpretation of the evidence that is gathered. Nondestructive tests may be for faults and discontinuities located on either the surface or internally and may be performed before, during, and after the manufacturing process.

These tests are performed by (1) exposing the product material to some kind of probing medium (radiation energy, sonic energy, magnetic and electrical energy, and other media), (2) obtaining some kind of indicating signals from the probing medium, and then (3) interpreting the signals as evidence of the presence or absence of possible defects. To function properly, a suitable probing medium must be one that can be applied in such a manner that it will be affected by any defects present, and the signals obtained must be correlated with the defects.

Standardized Tests. Over the years a number of tests have been standardized for checking of material properties. Some of these provide data that are useful for design calculation, while others have the primary purpose of aiding in material choices by supplying comparative information. Many properties are defined only by the test procedure that has been developed for their measurement. To cover the wide range of values occurring with different materials, shapes, and sizes, different sets of conditions have been established for some of the tests. For any test for which this is true, it is necessary that the test conditions used be indicated as part of the measurement.

THE TENSILE TEST

One of the more important tests for determination of mechanical properties of materials is the tension test. Material specimens are fastened between a fixed table and a movable table on a machine designed specifically for this purpose (Figure 3-4). A weighing scale is attached to the tables so that as they are moved apart (together for compression testing), the load imposed on the specimen can be measured. Some machines are fitted with auxiliary equipment that takes into account the loads imposed and the resulting elongation of the specimen to actually plot a stress-strain diagram of the test. The same results can be accomplished without this special equipment by measuring the elongation as the loads are increased and plotting the individual points to develop the curve.

Tensile Specimens. In order that these standard tests can be accurately reproducible and valuable for comparison with other tests, test specimens are made to one of several standard designs. Figure 3-5 shows the dimensions for a standard tension test bar with 8-inch gage length for rolled, flat stock. The radii from outside the gage-length portion to the increased section size at the ends are designed, in this and other test bars, to minimize stress direction effects from clamping loads on the end of the bar. Round test bars with the same 8-inch gage length are standard for testing rod and bar materials, but because it is often impossible to produce test samples of this length from castings and forgings and other material sources, a 2-inch gage length is frequently used. The diameter of the parallel section of round, tensile test bars is made to 0.505 inch (0.2 square inch cross-sectional area) to facilitate calculations. Adoption and use of the international metric system of measurement require that these dimensions be expressed in centimeters.

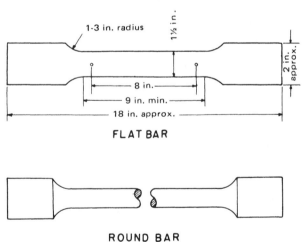

Figure 3-5
Tension test bars 8-inch gage length

Figure 3-4
Universal testing machine
Setup as shown for compression test

Stress-Strain Diagram. An understanding of a tensile test can best be acquired from a stress-strain diagram made by plotting the unit tensile stress against the unit strain (elongation), as shown in Figure 3-6. The illustration displays data from a tensile test on ductile steel and is representative of this kind of material only. Curves for other materials take on slightly different shapes.

Elastic Deformation and Plastic Flow. The straight line from *A* to *B* represents loads and deformations in the elastic range, and as long as the load at B is not exceeded, the material will resume its original position and shape after removal of the load. B is the elastic limit for this particular material, and loads above that limit will cause permanent deformation (plastic flow) that cannot be recovered by removal of the load. At the load represented by the point at C, plastic flow is occurring at such a rate that stresses are being relieved faster than they are formed, and strain increases with no additional, or even with a reduction of, stress. The unit stress at C is known as the yield point.

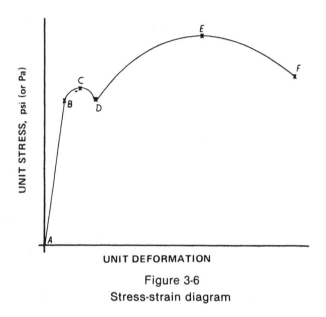

Figure 3-6
Stress-strain diagram

Plastic flow occurring at normal temperature is called cold working, regardless of the kind of loading system under which it is accomplished. As plastic flow takes place, the crystals and atoms of the material rearrange internally to take stronger positions resisting further change. The material becomes stronger and harder and is said to be work hardened. At the point D in Figure 3-6, the curve suddenly turns upward, indicating that the material has become stronger because of work hardening and that higher loads are required to continue deformation. The deformation rate, however, increases until at point E the ultimate strength is indicated.

Ultimate and Breaking Strengths. The ultimate tensile strength of a material is defined as being the highest strength in pounds per square inch, based on the original cross-sectional area. By this definition, ductile materials that elongate appreciably and neck down with considerable reduction of cross-sectional area, rupture at a load lower than that passed through previous to fracture. The breaking strength, or rupture strength, for this material is shown at F, considerably below the ultimate strength. This is typical of ductile materials, but as materials become less ductile, the ultimate strength and the breaking strength get closer and closer together until there is no detectable difference.

Yield Point and Yield Strength. Many materials do not have a well-defined or reproducible yield point. Plotting of tensile stress-strain values produces a curve of the type shown in Figure 3-7. For these materials, an artificial value similar to the yield point, called yield strength, may be calculated. The yield strength is defined as the amount of stress required to produce a predetermined amount of permanent strain. A commonly used strain or deformation is 0.002 inch per inch, or 0.2% offset, which must be necessarily indicated with the yield strength value. The yield strength is the stress value indicated by the intersection point between the stress-strain curve and the offset line drawn parallel to the straight portion of the curve.

Modulus of Elasticity. In the stress range below the elastic limit, the ratio of unit stress to unit deformation, or the slope of the curve, is referred to as the

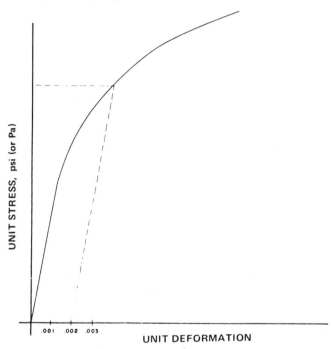

Figure 3-7
Yield strength

modulus of elasticity, or Young's modulus, and is represented by E. E, therefore, equals s divided by δ. Following are listed the values of E for some of the more common structural materials:

TABLE 3-1

Aluminum alloys	10 million psi (6.9 × 10⁹ Pa)
Copper alloys	14 to 19 million psi
Gray iron	12 to 19 million psi
Steel and high-strength irons	28 to 30 million psi
Cemented carbides	approx 50 million psi

The gross values of the modulus of elasticity are important to the design of members when deflection or deformation in the elastic range must be given consideration. The relative stiffness or rigidity of different materials can be ascertained merely by comparing their moduli. By rearrangement of the formula for E, the unit deformation becomes equal to the unit stress divided by E. If a bar of steel with a cross-sectional area of 1 square inch and with a modulus of elasticity of 30 million pounds per square inch is subjected to a tensile pull of 1,000 pounds, each inch of length of the bar will be stretched 1/30,000 of an inch. A 30-inch-long steel bar with this cross section would then be elongated 1/1,000 (0.001) of an inch overall with a 1,000-pound tensile load.

Ductility. The tension test provides two measures of ductility. One is called percent elongation, represented by the formula

$$\text{percent elongation} = \frac{(L_f - L_o)}{L_o} \times 100$$

where
L_f = final gage length
L_o = original gage length.

For ductile material the major portion of the elongation will occur over a relatively small portion of the gage length after the specimen begins to neck as it approaches the breaking point. Because much of the elongation is localized, a variation of gage length would cause a difference in calculated percent elongation.

Another measure, percent reduction of area, is calculated by comparing the original area of the specimen to the smallest area of the neck at rupture.

Resilience and Toughness. The area under a curve is influenced by both factors that are used to make that curve. In a stress–strain diagram the area under any portion of the curve represents the energy required to deform the material. Up to the elastic limit, this energy is recoverable and is called *resilience*. *Toughness* is defined as the ability of a material to absorb energy without fracture. For the tension test, the total area under the curve is a measure of toughness.

TRUE STRESS—TRUE STRAIN

In the tensile test just described, stresses were calculated as though the original specimen size did not change. More precisely, the vertical axis of the diagram should be labeled load/original area rather than stress. If each time a load reading were made, the smallest diameter of the specimen were found and the calculation for stress based on this actual diameter, this axis could be labeled true stress. The definition of true strain is somewhat more complex, and, in any case, true strain does not differ greatly from elongation normally plotted. The greatest difference between the diagram of Figure 3-6 and a true stress-flow strain diagram would be in the plastic flow region.

True stress would continue to increase throughout the test, as shown in Figure 3-8, and maximum stress would occur at the final break. The test of Figure 3-6 is usually called an engineer's stress-strain diagram. This curve is shown as a dotted line in Figure 3-8. Not only is it easier to prepare than a true stress-true strain diagram, but the value for ultimate strength obtained from it is more useful for design than the maximum true stress that occurs when the specimen breaks. The true concern of a designer is the maximum load that can be supported, not the maximum stress.

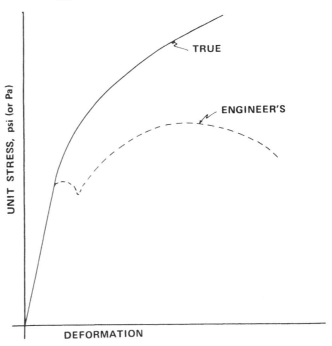

Figure 3-8
True stress-true strain diagram

COMPRESSION TESTING

Up to the elastic limit, most metals are approximately equal in properties under either tensile or compressive loading. Cast iron, however, has a tensile strength of only about one-half its compressive strength and is therefore used mostly in applications where the principal loads are of the compressive type. Many nonmetals such as timber, concrete, and other aggregates are also used almost entirely for supporting compressive or compactive loads. This is due in part to higher compressive strength, but also these materials have a high incidence of flaws and faults that might cause sudden failure in tension but produce relatively small effect under compressive loading.

The testing of materials in compression is conducted in much the same manner as in testing under tension. Specimens are placed between tables of a testing machine that are brought together to subject the specimen to compressive loads. Compression specimens must be short compared to their diameter so that column effect will not cause bending with eccentric, unequal loading.

TRANSVERSE RUPTURE TESTING

Limitations of Tensile Tests for Brittle Materials. In a number of cases a substitute for the standard tensile test is necessary. With some materials that are difficult to shape or very brittle in nature, it is impractical to produce a specimen for tension testing. This condition occurs particularly with ceramics. With most materials that are very brittle in character, even though a tensile specimen might be produced, the results from the standard tensile test would have only limited significance. It is almost impossible to insure in the tension test that the applied load will be precisely centered in the specimen and will be exactly parallel to the axis of the specimen. If this is not the case, bending moments are introduced in the specimen. With a ductile material, small amounts of plastic flow take place in the specimen, particularly where the load is applied; the specimen aligns itself properly with the load; and the stresses are uniform across the tested area. With a brittle material in which this alignment cannot take place, the bending moments result in higher stresses on one side of the specimen than on the other. The specimen fails when the highest stress reaches some critical value, but the observed stress at this time, based on the assumption of uniformity, is somewhat lower. As a consequence, the results from testing a number of similar brittle specimens exhibit wide variations and are not representative of the true strength of the material.

The Transverse Rupture Test. The transverse rupture test, while it gives less complete information than the tension test, is a fast and simple test, making use of more easily prepared specimens, and is especially well suited to brittle materials. In many instances the specimen can be an actual workpiece. The test is particularly well suited for those materials that are to be used in beam applications. It is really the only meaningful type of strength test for reinforced concrete.

The test consists of loading a simple beam as illustrated in Figure 3-9. While some standards have been set for particular materials, there are no univeral standards for specimen sizes and shapes as there are for the tension test.

The modulus of rupture, or beam strength, is calculated by the formula

$$S_r = \frac{3PL}{2bd^2}$$

Limitations of Transverse Rupture Testing. While this formula is the formula that is used to calculate the maximum actual stress in the outer fibers in a beam, it is based on the assumption that stress remains proportional to strain. This is not the case for most materials when highly loaded, with the result that the calculated "stress" is higher than the actual stress in the outer fibres at rupture, and direct comparison cannot be made with ultimate tensile strength values taken from a tension test, nor can the values of modulus of rupture be used as design tensile strength values. The values are useful for comparing materials, and they are useful in design when the material is to be used as a beam.

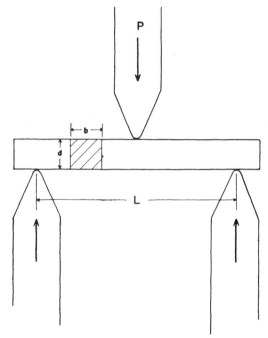

Figure 3-9
Transverse rupture test

SHEAR TESTING

In the section dealing with material failure, it was pointed out that when a bar is subjected to a tension load as in the tension test, the value of shear stress existing in the bar at failure can be calculated from the load and the dimensions of the bar (Figure 3-10).

Figure 3-10
Heads of a torsion testing machine. Torsion is the simplest way of obtaining pure shear stress. Results are useful for evaluating cold-working properties of metals

The term *shear*, however, has a broader meaning than shear stress only and is used to describe loading systems that subject a material to a shearing action. Actually, the stress distribution in such loading system is quite complex, but a rather simple *shear strength* test has been developed that simulates the conditions of actual loading and provides information that may be used in deisgn where the loading situation is similar to that of the test. Such loading occurs in using bolts or rivets and in shearing operations in which material is being separated. In the test indicated in Figure 3-11, the bar with cross-sectional area A is made to fail simultaneously in two places so that the area of failure is 2A, and shear strength is defined as shear strength = $P/2A$.

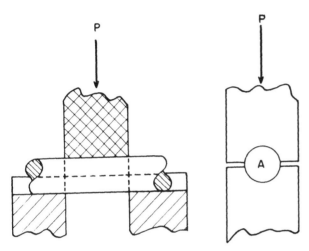

Figure 3-11
Shear strength test

FATIGUE TESTING

A metal may fail under sufficient cycles of repeated stress, even though the maximum stress applied is considerably less than the strength of the material determined by static test. Failure will occur at a lower stress level if the cyclic loading is reversed, alternating tension and compression, than if the cycles are repeated in the same direction time after time. The conclusion from one comprehensive study of service failures was that in 90% of such failures in which fracture occurred, fatigue was involved. Structural members subject to vibration, repeated variation of load, or any cyclic disturbance causing deflection must be designed to have low enough stress levels that fatigue phenomena will not cause failure.

Fatigue Failure Initiation and Development. Fatigue failure normally starts at some spot where stress concentration is high because of the shape of the member or some imperfection. Holes through the material, notches in the surface, internal flaws, such as voids, cracks, or inclusions or even minor scratches and faults caused by corrosive attack on the grain boundaries, may be sources of fatigue failure. With repeated stressing, a crack starts at one of these fatigue nuclei and grows until insufficient solid metal remains to carry the load. Complete failure in a sudden, brittle manner results. As seen in Figure 3-11, the exposed surface of a fatigue failure shows part of the surface to be smooth and polished, while the rest exhibits a well-defined grain structure. The crystal-line-appearing portion was separated in the sudden, final break. The smooth part was polished and burnished by the movement of the material with repeated deflection as the crack developed and grew.

Fatigue failure is more frequent than commonly thought. There have been estimates that with equipment having moving parts or subject to vibration as much as 90% failures include fatigue in some form. Because any kind of discontinuity, particularly those at (or near) the surface where tensile stresses are likely to be highest, can be the nucleus for fatigue failure, location of these spots by NDT may prevent a later catastrophic failure.

Endurance Limits. Because a material may fail under conditions of a great many repeated loads at a stress level far below that determined by the standard strength test, a designer must know how different materials stand up under these conditions. Tests have been developed with special machines that bend plate-shaped test specimens or subject a rotating beam to a bending load for large numbers of cycles. From data collected from such tests, the endurance limit of a material can be determined.

The *endurance limit* is the highest completely reversed stress whose repeated application can be endured for an indefinitely large number of cycles without failure. Figure 3-12 shows a typical *S-N*, or endurance limit, curve. The material represented by this curve would have an endurance limit of 42,000 pounds per square inch (290 MPa) because the curve

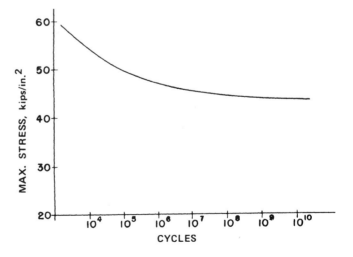

Figure 3-12
Typical *S-N* curve

has flattened out, and stressing at this level could be continued indefinitely without failure. Endurance limits correlate fairly closely with tensile strength and for most materials are from about one-third to one-half the stress required to break a tensile specimen.

Fatigue Strength. For some materials the curve does not flatten even after several hundred million cycles. When the endurance limit cannot be determined, or it is impractical to carry on a test long enough for this determination, it is common practice to use another value, fatigue strength, to evaluate the ability of a material to resist fatigue failure. *Fatigue strength* is the stress that can be applied for some arbitrary number of cycles without failure. The number of cycles for which a fatigue strength is valid must always be specified because the operating stress chosen may be at a level where the *S-N* curve still slopes, and indefinite cyclic operation could cause fatigue failure.

CREEP TESTING

The term *creep* is used to describe the continuous deformation of a material under constant load, producing unit stresses below those of the elastic limit. At normal temperature, the effect of creep is very small and can be neglected. As operating temperatures increase, however, this deformation by slow plastic flow becomes very important in the design and use of material. Recognition of this phenomenon is most important for the higher strength materials that are to be used at elevated temperatures.

Creep tests are conducted by applying a constant load to a material specimen held at the desired temperature and measured periodically for deformation over a long period of time. The results may be plotted on a graph of elongation against time, as in Figure 3-13, with an indication of the maintained temperature and stress level under which the test was conducted. Most creep tests are carried on for periods of at least 1,000 hours, so this is a time-consuming test. The *creep strength* of a material is the stress required to produce some predetermined creep rate (the slope of the straight portion of a curve) for a prolonged period of time. Commonly, the stress required to produce a creep rate of 1% in 10,000 hours is used as creep strength. *Stress rupture strength* is defined as the stress required to produce failure at prescribed values of time and temperature.

NOTCHED BAR TESTING

Materials are often used in situations in which dynamic loads are suddenly applied to produce shock that increases the effective load far above that which would be expected from gradual application of the same load or a similar static load. Tests designed to check the ability of a material to withstand this kind of loading are energy absorption tests that seldom can

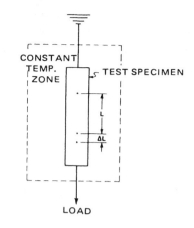

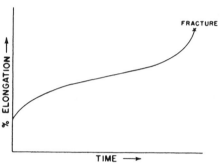

Figure 3-13
Creep test

be used to give information that can be used directly in design, but primarily provide data for comparison of different materials. While such tests are frequently called impact tests, the energy required to cause failure does not differ greatly from that required if the load were applied slowly. True impact failure, in which the energy-absorbing capacity of a material is greatly reduced, occurs only at much higher speeds.

Charpy Test. The most commonly conducted tests are bending impact tests, using one of two kinds of notched speciments (Figure 3-14). The Charpy specimen is supported at both ends by a standard

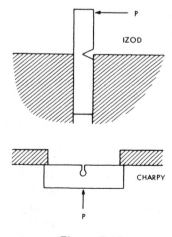

Figure 3-14
Impact specimens

impact testing machine and struck on the side opposite that of the notch. The testing machine is constructed with a weighted pendulum, which is lifted to start the test. Upon its release, the pendulum swings past the specimen, and breaks it. As the pendulum swings past, the remaining energy can be measured by the height of the swing and the absorbed energy determined.

Izod Test. The Izod specimen is supported in the testing machine by one end only and is loaded as a cantilever beam with a notch on the side of impact. Energy absorption is measured in the same way as with the Charpy specimen.

Test Specimens. Two kinds of notches are used on bending impact specimens. The Izod specimen is usually made with a 45° angular notch with a 0.010-inch radius at the bottom. The specimen is extremely sensitive to variation of notch size or change of radius, and extreme care in manufacture of the test specimen is necessary for reproducibility of test results. The keyhole notch shown on the Charpy specimen can be duplicated more accurately but is limited in the smallness of the hole producing the notch effect by the size of the smallest drill that will not "drift" in making the hole. The notches in the test specimens act as points of stress concentration, and the smaller the notch radius, the more severe is the stressing at this point. These notched test specimens actually provide only information regarding material that is to be used in a similar notched condition but are often practical because materials are frequently used with design shapes or structural imperfections that cause a structural member to be, in effect, a notched beam.

Tensile Impact Test. Greater reproducibility and greater similarity between the test and some use conditions can be provided by tensile impact tests. The specimens for these tests are not notched and are supported so that uniaxial tensile impact loads may be applied. The standard impact testing machine with pendulum weight can be tooled for testing small specimens of this type. For larger specimens a special machine with a variable-speed flywheel to store energy can be obtained.

BEND TESTING

Materials that are to be deformation processed by being subjected to bending loads and materials that may have been affected by localized heating, such as in welding, are sometimes tested by bend tests to provide comparative data.

Free Bend Test. Free bends are accomplished by prebending a flat specimen slightly to produce eccentricity and then loading the specimen in compression (column) until failure occurs or a 180° bend is produced. Normally, the loads to accomplish this are so variable that they are of little value and are not recorded. Instead, the angle of bend at failure is compared with results of other tests.

Guided Bend Test. In guided bend tests, the test specimen is bent about a fixed radius to 180°. The bend angle of a failure before 180° bending usually cannot be satisfactorily compared with other test results because of nonuniform plastic flow of material in the specimen caused by pressures set up by the guided bend fixture. Multiple-radius guided bends may be used for rating specimens by determining the smallest radius about which a standard specimen will bend 180°.

HARDNESS TESTING

The most frequently used tests for determining material properties are hardness tests. With sufficient knowledge of material composition and previous processing, hardness tests can be used as indirect measures of properties entirely different from hardness. For example, hardness can sometimes be used to separate raw materials of different composition, to determine whether or not satisfactory heat treating or other processing has been accomplished, or to measure the strength and wear-resistant properties of a product. Hardness measurements, therefore, are frequently made on raw material, on parts in process, and on finished goods ready for use.

With some metal alloys, electrical conductivity and hardness are related within limited ranges. Eddy current tests standardized to measure electrical conductivity can therefore be used as an indirect measure of hardness. Such tests must be applied cautiously since the ranges are restricted over which the relationship between hardness and conductivity are reasonably linear. Aluminum alloys and other non-ferrous metals are more reliably tested by this method than are ferrous alloys.

Most hardness tests result in some kind of measure of the ability of a material to resist penetration of the near surface material. Penetration of material with any kind of indentor requires the use of force and involves plastic flow of the tested material. The work-hardening qualities of a material, therefore, become part of most hardness measurements and partially explain the difficulty of converting from one type of hardness measure to another, because different methods of measuring hardness do not measure exactly the same thing. They are, however, well enough standardized to provide useful and practical information.

Mohs Test. One of the first standardized systems of measuring hardness made use of the *Mohs* scale of hardness, which specifies ten standard minerals arranged in order of their increasing hardness and numbered according to their position. Starting with number 1 as the softest, the standard Mohs scale is as follows:

1 ... Talc
2 ... Gypsum
3 ... Calcite
4 ... Fluorite
5 ... Apatite
6 ... Orthoclase (Feldspar)
7 ... Quartz
8 ... Topaz
9 ... Corundum
10 ... Diamond

If a material can be noticeably scratched by the mineral topaz (number 8) but cannot be scratched by quartz (number 7), it would have a hardness value between 7 and 8 on the Mohs scale. The Mohs scale of hardness has little value for hardness testing of metals but is still widely used in the field of mineralogy.

File Test. Another abrasion or scratch method of measuring hardness that does have some practical use in metal working is the file test. Standard test files can be used to gage quickly the approximate hardness of a material and, although not very accurate, can be used in many shop situations with satisfactory results. Experience and comparison with standard test blocks will permit a fair degree of accuracy to be attained.

Brinell Test. In 1900 Johan August Brinell, a Swedish engineer, introduced a new universal system for hardness measurement. The method involves impressing, with a definite load, a hardened steel ball into the material to be tested and calculating a *Brinell* hardness number from the impression size (Figure 3-15). A wide range of hardnesses can be tested by varying the size of the ball and the loads imposed, but in the hardness range most frequently tested, a ball 10 millimeters in diameter is impressed into the material under a load of 3,000 kilograms for 10 seconds to check steel and under a load of 500 kilograms for 30 seconds to check nonferrous materials. The numerical value of the Brinell hardness number is obtained by dividing the load in kilograms by the area of the spherical impression in millimeters. In practice, the average diameter of the impression is usually read with a measuring microscope and the Brinell hardness number determined directly from a table.

Advantage and Limitation of Brinell Tests. The Brinell hardness method has the advantage, as compared to most other measuring methods, of determining a hardness value over a relatively large area, thus reducing the inconsistencies caused by flaws, imperfections, and nonhomogeneity in the material, likely to be introduced with small area measurement that includes only a few metallic grains. With plain carbon and low alloy steels, the relation between tensile strength and Brinell hardness is so consistent in the medium hardness range that the tensile strength of the steel can be closely approximated by multiplying the Brinell hardness number (BHN) by 500. The principal disadvantages of the Brinell method are that the machine to supply the load for impressing the ball into the material is often cumbersome and cannot always produce the impression where desired. The ball cannot be impressed in very thin materials and, of course, cannot be used to examine extremely small samples, and the impression is of such size that it may harm the appearance or use of finished surfaces.

Rockwell Test. Because of its convenience and the fact that only small marks are left in the work tested, one of the most frequently used tests is the *Rockwell* hardness test (Figure 3-16). This also is an impression test, but the hardness number is determined by a differential depth measurement that can be read directly on a dial indicator of the machine used to impose the load (Figure 3-17 shows the Rockwell hardness tester.) To obtain a Rockwell

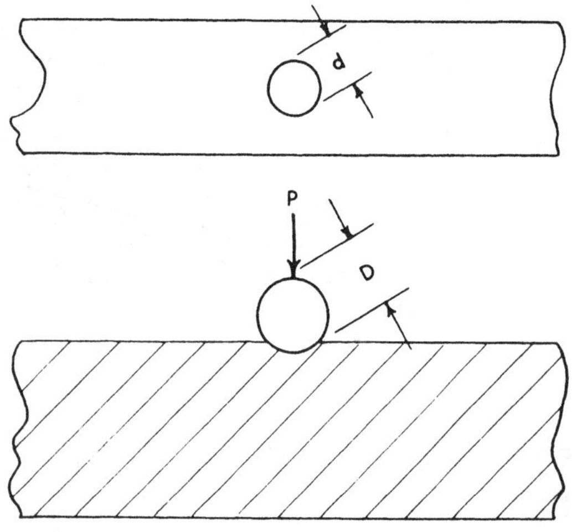

$$BHN = P / \frac{\pi D}{2}(D - \sqrt{D^2 - d^2})$$

Figure 3-15
Brinell hardness measurement

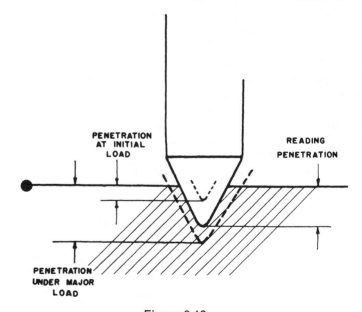

Figure 3-16
Rockwell hardness measurement

hardness reading, the equipment is first used to place a minor load of 10 kilograms on the penetrator. This reduces the effect of dirt, oil films, scale, and other surface conditions that might affect the reading. A major load of 60, 100, or 150 kilograms, depending upon the type of penetrator and scale being used, is then imposed to force the penetrator into the work material. After the penetrator has seated to its full depth — the time usually being controlled by a dash pot built into the equipment — the major load is removed. The permanent differential depth between the minor and major loads is then read directly as a Rockwell hardness number.

Standard Rockwell Scales. Although provision has been made for use of a 1/8-inch-diameter ball as a penetrator, almost all hardness testing with the Rockwell equipment is done with two standard penetrators. The one used for softer materials is a 1/16-inch-diameter hardened steel ball supported in a

Figure 3-17
Rockwell hardness tester. The tester impresses a penetrator into the work to provide a direct surface hardness measurement

special chuck that permits easy replacement should the ball become damaged. The testing of harder materials that would cause excessive deformation of the hardened steel ball is performed with a diamond-tipped penetrator with a 120° conical point and a spherical tip of 0.200-millimeter radius. The diamond penetrator, or indentor, is known as a brale.

The penetrator used and the size of load impressing it into the test material are defined by a letter that becomes part of the Rockwell reading. The accompanying Table 3-2 shows the relationship among the scale designation, the loads, and the penetrators.

TABLE 3-2

Scale	Load kilograms	Penetrator
A	60	Brale
B	100	1/16" ball
C	150	Brale
D	100	Brale
F	60	1/16" ball
G	50	1/16" ball

The letter designating the test conditions is a very important part of a hardness notation because the number alone could represent several different hardness conditions. For example, a Rockwell hardness reading of B 60 would represent a relatively soft material, such as a medium hard copper alloy. A Rockwell hardness reading of C 60, sometimes written R_C 60, on the other hand would represent a hardness such as might be used for a hardened tool steel to cut metals.

Superficial Rockwell Test. Another machine, the Rockwell *superficial* hardness tester, is contructed and used in much the same manner as the standard machine but is a special-purpose tester designed to be used when only a very shallow impression is permissible or when measurement of hardness of material very close to the surface is the principal aim. The superficial hardness tester makes use of the same penetrators, except that the brale is of higher precision and is designated as N brale. The loads used to cause penetration are lighter: 15, 30, and 45 kilograms. Table 3-3 shows the testing conditions for Rockwell superficial hardness testing.

TABLE 3-3

Scale	Load kilograms	Penetrator
15N	15	N brale
30N	30	N brale
45N	45	N brale
15T	15	1/16" ball
30T	30	1/16" ball
45T	45	1/16" ball

As in the previous case, the scale indication must be used as a prefix to the hardness number read from the dial.

Vickers Test. The *Vickers* hardness tester operates on the same principle as the Brinell instrument but makes use of a diamond penetrator shaped as a four-sided pyramid. The impression made by the penetrator is accurately measured by swinging a microscope into position without moving the test piece in the machine. As in the Brinell method, the Vickers hardness number is the ratio of the force imposed on the indentor to the area of the pyramidal impression. In the lower range of hardness, under Brinell 300, Vickers and Brinell hardness numbers are almost identical, but above this range they separate as hardness increases, primarily because of distortion of the steel ball used for Brinell testing when it is forced against the harder materials.

Microhardness. It is frequently important, particularly in research or development work, to test the hardness of material that is very thin or very small in area. A number of special machines have been developed for determining "microhardness." One of the more commonly used pieces of equipment of this type is the *Tukon* microhardness tester. Normally, the machine is fitted with an elongated diamond-shaped penetrator. Microscopic measurement of the impression provides information that can be converted to *Knoop* numbers. Knoop hardness measurement often cannot be compared directly with Brinell or Vickers hardness measurement because the elongated impression is rather strongly affected by the directional properties of the material being tested. The use of a symmetrical, square-based, pyramid-shaped indentor will provide hardness data comparable with that of the other systems.

It should be self-evident that the lighter the indentor loads and the smaller the impressions made, the greater the care that must be used to perform a hardness test, and the better must be the quality of surface on which it is made. In Brinell testing, small surface imperfections tend to be averaged out because of the large area covered, but in microhardness checks, in which the impression may be only a few thousandths of an inch long, small scratches and surface imperfections may contribute large errors. Microhardness testing is usually performed on a highly polished surface, and in many cases, to obtain reproducibility, it is necessary to etch the surface to reveal the constituent structure in order to locate the impression properly.

FACTOR OF SAFETY

No property, structural or otherwise, whether calculated from theoretical considerations or determined by test procedures, can be safely used at or very close to its ultimate (maximum) value. Tests are neither consistent enough nor accurate enough, particularly as they are not conducted under exact use conditions, to permit strong confidence to be placed in their results. Also, because of the complexity of stress-analysis problems, it is almost essential that simplifying assumptions be made during design to prevent design costs and time from becoming prohibitive. A factor of safety is therefore used to prevent working too close to maximum values. The factor of safety is the ratio between the maximum value and the working value and is determined by competent judgment, taking into consideration all conditions of use. Factors of safety vary from as low as one to as high as five or more. They may be applied to any quality but are most commonly used in connection with strengths.

As an example of its use, if the ultimate tensile strength of a certain grade of steel is 80,000 pounds per square inch and its elastic limit, 60,000 pounds per square inch, an allowable stress, or working stress, of 20,000 pounds per square inch would provide a safety factor of four, based on the ultimate strength, or of three, based on the elastic limit.

The closer the factor of safety approaches one, the more the danger that an unforeseen fault or condition of use may cause failure. On the other hand, the larger the factor of safety, the greater the volume and weight of material needed, with a corresponding increase in cost and in space-need problems. Factors of safety in the range of two to four are most common, but a satisfactory value depends upon a great number of conditions, some of which are described in the following paragraphs.

Allowances must be made for unexpected loads or conditions. This is particularly true if the human element is large in the use of the equipment, since the human mind is most unpredictable. It is common to include a factor of at least two in the factor of safety when a design is based on static tensile strength values but subjected in use to varying loads. This corresponds approximately to the ratio of static tensile strength to endurance limit.

Allowances must be made for environmental and time factors. Strengths of most materials are greatly reduced by corrosion and other chemical effects. Other materials lose strength or become brittle with age. The consistency of test data should influence the factor of safety choice. Test information should be of large enough volume to be statistically significant. Larger safety factors are necessary with materials varying widely in quality than with those that are quite uniform.

Whether or not the use of a material may affect human life has a large influence on the factor of safety. In the designing of hoists, cranes, and other lifting equipment, factors of safety of five or more are commonly used because failure could mean injury or loss of life. The same consideration applies, of course, to aircraft design. Here, however, space and weight are very important, and large factors of safety could easily prevent a usable design; consequently, the problem is handled in a different way. Extreme care is used in selecting and testing materials. Stresses are carefully calculated and, as far as possible, the structures built so that they cannot be overloaded in use. Thus, by spending more care, time, and money preceding and during manufacturing, it is possible to use a smaller factor of safety because of greater certainty of not exceeding the design condition. The smaller the safety factor and the more important any possible failure, the more reliable must be any nondestructive testing procedure that is used.

The Nature of Materials and Solid State Changes in Metals   4

The chemist ordinarily considers the smallest functional portion of matter to be the atom. The atom consists of a nucleus, made up of positively charged protons and uncharged neutrons, surrounded by electrons. The electrons carry negative charges and move in orbit at different levels. Each level of orbit can contain only a definite number of electrons, and the number of levels or shells is determined by the atomic number of the element. All the shells will usually be full except the outer one, which is short of the maximum possible number of electrons for most materials. All of the electrons are in constant motion, spinning about their own axes and traveling through their orbits about the nucleus with speeds dependent on their energy level, which in turn is strongly affected by the pressure and the temperature conditions. The physicist's picture of an atom depicts it as a heavy nucleus containing most of the mass, surrounded by a cloud of moving electrons.

THE EFFECT OF ENERGY ON THE ATOM

Forces on the Atom. A number of different forces exist among the atoms making up a material, some of them attractive, some repulsive. The nature of any material depends primarily on the nature of these forces, which themselves depend not only on the type of atom, but also on the energy level of the atom. At high energy levels, the repelling forces predominate, and the atoms tend to move as far from each other as possible. This condition is called the *gaseous* state. If the energy of the material is lowered, the forces change, and a condition of equilibrium is reached in which the atoms assume fixed average distances from each other, although still free to move and not tied closely together. In this *liquid* state, the materials have fixed volume but assume the shape of the container in which they are placed.

As the energy level is further decreased, the mobility of the atoms decreases. There are at least four different mechanisms by which the atoms can assume positions well fixed enough that for practical purposes the material could be called *solid*. Of the materials of interest to manufacturing, all the metals occur as *crystalline* solids.

METALLIC STRUCTURE

Definition of a Metal. Metals are usually defined as materials having some degree of plasticity, relatively high hardness and strength, good electric and thermal conductivity, crystallinity when solids, and opacity. A definition based on atomic structure is more precise. A metallic solid is one that has free electrons available in the structure to carry a current and that has a negative coefficient of conductivity with increasing temperature.

States of Matter. Figure 4-1 shows the relationship that exists among the three states of matter for a crystalline material. At the intersection of temperature T_1 and pressure P_1 on the curve, notice that an increase of temperature of a material for which this curve is valid would cause the material to change directly from a solid to a gas. Similarly, a reduction of pressure (a shift toward the left) would also cause the same change. Such a change of state from solid directly to gas is known as *sublimation*. Arsenic is the only metallic material that sublimates at atmospheric pressure. When the temperature is raised to T_2 at pressure P_2, the atoms of the material will become sufficiently active that a change is made from a solid to a liquid. A further increase in temperature at this same pressure to point T_3 will cause a second change from a liquid to a gas. The intersecting point of the curves at the temperature T_x and pressure P_x is known as the *triple point* and occurs at the temperature and pressure conditions under which a material may exist as a solid, a liquid, a gas, or partially all three at the same time. For most metals, this point occurs below normal temperatures and well below atmospheric pressure; consequently, most metals upon being heated go through the changes from solid to liquid to gas as the temperature increases.

Space Lattices. As the energy of a liquid metal is reduced by taking away heat, the attraction between atoms increases until they arrange themselves in definite three-dimensional geometric patterns that are characteristic of the metal. These structures are called *space lattices* and consist of network groupings of identical *unit cells* that are aligned in parallel planes.

There are fourteen types of crystal lattices, but most of the common and commercially important metals exist, in the solid state, in one of three structures. These are, as shown in Figure 4-2, body-centered cubic, face-centered cubic, and hexagonal closed-packed. In the illustrations of unit cells, the dots representing atoms should be considered as centers of activity for the atoms and not as graphic illsutrations of the atoms themselves.

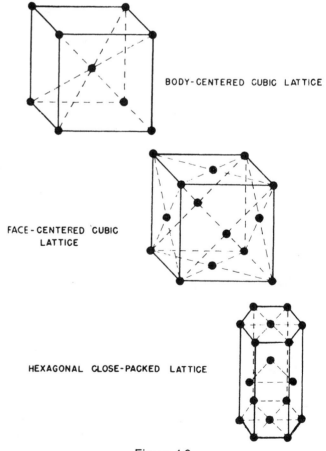

Figure 4-2
Common metallic space lattice

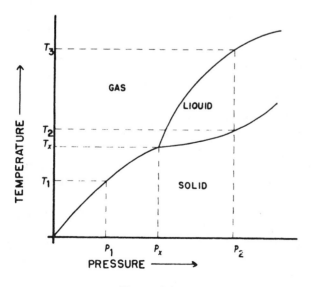

Figure 4-1
States of matter

A single unit cell does not exist alone. To attain stability, it must grow past some critical size by being joined with other cells that share the atoms on the outer adjacent surface. For purposes of illustration, it has been assumed that a unit cell can exist by itself and that all its atoms belong to it alone.

Body-centered Cubic Lattice. The body-centered cubic cell is made up of nine atoms. Eight are located on the corners of the cube with the ninth positioned centrally between them. The body-centered cubic is a strong stucture, and in general, the metals that are hard and strong are in this form at normal temperatures. These metals include chromium, iron, molybdenum, tantalum, tungsten, and vanadium.

Face-centered Cubic Lattice. Face-centered cubic cells consist of fourteen atoms with eight at the corners and the other six centered in the cube faces. This structure is characteristic of ductile metals, which include aluminum, copper, gold, lead, nickel, platinum, and silver. Iron, which is body-centered cubic at room temperature, is also of the face-centered structure in the temperature range from about 910° C to 1,400° C. This is a solid-state change that will be discussed more thoroughly in the following chapter.

Hexagonal Close-packed Lattice. Seventeen atoms combine to make the hexagonal close-packed unit cell. Seven atoms are located in each hexagonal face with one at each corner and the seventh in the center. The three remaining atoms take up a triangular position in the center of the cell equidistant from the two faces. The metals with this structure are quite susceptible to work-hardening, which will be discussed in the following chapter. Some of the more commonly used metals that crystallize with this structure are cadmium, cobalt, magnesium, titanium, and zinc.

Tin is an exception to the other commonly used metals in that the atomic configuration is body-centered tetragonal, which is similar to the body-centered cubic but has wider atomic spacing and an elongated axis between two of the opposite faces.

SOLIDIFICATION

Growth of a Crystal. As the temperature of the liquid metal is reduced and the atoms become less active, they are attracted to each other and take definite positions to form unit cells. Because cooling cannot be exactly the same for every atom, certain ones will assume their positions ahead of others and become a nucleus for crystal formation. In the process of assuming their positions, these first atoms will give up kinetic energy in the form of heat, which retards the slowing down of other atoms; but as heat removal is continued, other atoms will take their places along the sides of the already solidified unit cell, forming new cells that share atoms with the first and with others to come later. Orderly growth continues in all directions until the crystal, or as usually referred to for metals, the *grain*, runs into interference from other grains that are forming simultaneously about other nuclei.

Although with some metals and with special treatments it is possible to grow single crystals several inches in diameter, with most metals and at the usual cooling rates, great numbers of crystals are nucleated and growing at one time with different orientations. If two grains that have the same orientation meet, they will join to form a larger grain, but if they are forming about different axes, the last atoms to solidify between the growing grains will be attracted to each and must assume compromise positions in an attempt to satisfy a double desire to join with each. These misplaced atoms are in layers about the grains and are known as *grain boundaries*. They are interruptions in the orderly arrangement of the space lattices and offer resistance to deformation of the metal. A fine-grained metal with large numbers of interruptions, therefore, will be harder and stronger than a coarse-grained metal of the same composition and condition.

Grain size, grain orientation, and the composition of grain boundaries are factors that can influence some nondestructive tests. In radiography, at certain x-ray energies, diffraction effects can produce images that resemble flaws and, at best, make interpretation difficult. Ultrasonic testing of large grained castings and welds also may be radically influenced by excessive noise and attenuation, which may produce false indications or mark the presence of dangerous flaws.

GRAIN SIZE

The grain (crystal) sizes produced during solidification are dependent both upon the rate of nucleation and upon the rate of growth of grains. For most materials the rate of growth is relatively slow, and the primary influence on grain size is the rate of nucleation. Grain size can be used as an indication, or measure, of properties. For this reason, visual standards have been set up to aid accurate comparisons. While not in routine usage, ultrasonic methods have been applied to grain size determination. If the grains are randomly oriented, at high ultrasonic frequencies, reflection from grain boundaries that would usually be considered noise can be related to grain size.

Importance of Grain Size. Grain size exerts an important influence on the mechanical properties of materials and, fortunately, can be controlled by methods much more precise than manipulation of the factors that influence growth during solidification. In some processes though, particularly casting, the solidification grain size is important, because with some materials and some shapes, grain size cannot be readily changed after the first formation. In those cases in which changes can be effected, additional processing costs will be added. The methods, other than solidification, that can be used for grain-size control involve solid-state changes.

As has already been indicated, coarse grains in the harder materials have lower strength than fine grains. Coarse-grained materials machine more easily, requiring less power, although the quality of surface produced will not be as good as with a finer-grained material. Coarse-grained ferrous material is easier to harden by heat treatment than fine-grained material of the same composition but has increased susceptibility to cracking under the thermal loads. Coarse-grained material will caseharden on the surface more readily than fine-grained. It is evident, then, that coarse grains may sometimes be desirable during processing, but fine grains are usually necessary in the final product to provide the best mechanical properties. Some deformation processes of shaping materials can be used so as to cause grain-size reduction automatically during the shaping process with little or no additional cost involved.

SOLID STATE CHANGES IN METALS

In the previous section the process of metal solidification was briefly described. The properties of a material are derived from the crystalline structure, including the atomic arrangement and the crystal sizes, and are affected by the boundary layers that join the grains together. The atomic arrangement is primarily a function of the material composition, which may consist of a single material or a combination of materials that are completely soluble, partially soluble, or totally insoluble in each other in the solid state. The structure and grain size also may be influenced by the operating temperature changes and by mechanical loads that stress the material sufficiently to cause plastic flow in combination with time and heat effects.

Some materials, particularly those that are cast to shape, may be used with the structure in which they solidify, but some of the cast materials and nearly all metals processed by other methods are treated in some way in the solid state to obtain improved mechanical properties.

These treatments include work hardening, recrystalization, age hardening, and heat treating of allotropic materials to cause crystal transformations. In many cases, treatment may be inherent in the process. This may be beneficial, as in many cases of deformation shaping with associated work hardening, or may be detrimental, as in other cases in which cold working develops directional properties in a material to make some kinds of further cold work difficult or impossible.

WORK HARDENING

Effects of Deformation. The application of loads to a solid material in processing or in service can cause two kinds of deformation. If the load does not stress the material past its elastic limit, the deformation is "elastic," and the material returns to its original position upon removal of the load. If, however, the elastic limit is exceeded, the material does not return completely to its original position when the load is removed and is permanently deformed by plastic flow within its crystalline structure. When the elastic limit is passed, elastic properties are not lost, but instead are enhanced, providing the deformation is produced by cold work. The strength of metal is increased by plastic flow and the elastic limit is raised. Some of the deformation processes produce improved properties at the same time the shaping is being performed.

PLASTIC DEFORMATION

Permanent deformation of metallic crystals occurs in three ways: slip, twinning, and rotational deformation. The degree of each is dependent largely on the characteristics of the particular metal.

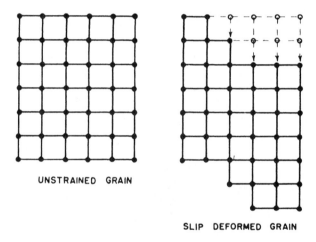

Figure 4-3
Slip

Slip Deformation. *Slip* deformation is illustrated in Figure 4-3 and occurs by translation or sliding between the atomic planes within a grain. If the deformation causes more than a very minor shift, a large number of atomic planes in each grain will slide over adjacent planes to occupy new locations with new neighbors. The planes through the crystal that are usually most subject to slip are those of the greatest atomic population and greatest distance between planes. The orientation of the planes along which slip takes place most easily will, of course, be different for different types of crystal lattices. Because of the usual random orientation of the crystals, the slip planes of many will not be in line with the direction of loading. When the best slip planes are completely out of alignment, slip may occur along other less preferred planes.

Twinning Deformation. Figure 4-4 shows a type of grain deformation referred to as *twinning*, which seems to occur most easily under loads applied suddenly, rather than gradually. With twinning, the grain

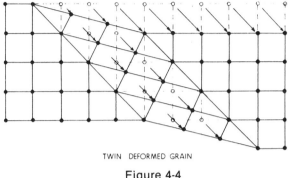

TWIN DEFORMED GRAIN

Figure 4-4
Twinning

deforms by twisting or reorienting a band of adjacent lattice forms, with each unit cell remaining in contact with the same neighbors it had before deformation took place.

Rotational Deformation. A third type of shift in a grain is a kind of rotational deformation of portions of the crystal lattice. Stresses below the elastic limit cause the crystals to be temporarily bent and deformed, but when the elastic limit has been exceeded and slip has occurred on a number of different planes, sections of the lattice tend to bend and rotate to a new, preferred orientation. After a large percentage of grains have been reoriented by action of considerable deformation work, the metal is likely to take on directional properties called *fibering*. Fibering may be beneficial or harmful, depending upon the use to which the material is put.

COLD WORK

According to dislocation theory, as plastic flow takes place, existing dislocations (atomic discontinuities) are reinforced and new dislocations are created to resist further plastic movement. Regardless of what the exact mechanisms may be by which plastic flow takes place in the metal grains, it is a proven fact that when metals are cold worked to produce plastic deformation, they become harder and stronger. The word *cold* in this instance refers to different temperatures for different metals. Cold work is work accomplished below the recrystallization temperatures for the particular material.

The mechanical strain energy necessary to produce the plastic deformations described above is converted to other forms of energy within the material. In most metals, dislocation processes are accompanied by discrete releases of mechanical energy, sometimes called stress waves. These stress waves produce acoustical vibrations that travel at high velocity through the material and can be detected by sensitive microphones or transducers coupled to the metal surface. Analysis of the intensity, duration, and rate of such acoustic emissions can provide information about the formation and growth of microcracks that result from continued loading. Acoustic emission monitoring is a relatively new means of nondestructive testing that has the capability of signaling states of over-stress and producing early warning of impending failure.

RECRYSTALLIZATION

Metals that are cold worked are left with their grains in a strained and unstable condition. The grains have a tendency to return to the equilibrium of a lower energy state by equalization of internal crystalline stress or by changing to new, unstrained grains. The greater the deformation strain, the greater the instability and the easier it is for the change to take place. Time and temperature also have strong influence. Two kinds of change, recovery and recrystallization, take place upon the heating of a cold-worked metal.

RECOVERY

First Effect. Recovery, sometimes referred to as stress relief, involves rearrangement of some of the more strenuous dislocations or imperfections with little or no effect on the external form of the crystals or grains. Although the changes that take place during recovery are rather minor in respect to the crystal, they have a marked effect on some properties. Electrical properties and corrosion resistance are improved and residual stresses are reduced.

Affected by Time, Temperature, Cold Work. Recovery occurs completely for some metals at room temperature. For some others it occurs partially over a long period of time without increase of temperature. For most it is necessary to heat treat to a specific temperature that will depend upon the degree of recovery desired. The temperature chosen will, of course, be dependent upon the metal and to some extent on the amount of cold work that has been performed previously. The objective of recovery is usually to regain electrical and chemical properties without sacrifice of mechanical properties. If the temperature is raised too high or maintained for too long a time, hardness and strength of the metal will decrease appreciably, but high temperature treatment is sometimes necessary to remove residual stresses in forgings and steel weldments.

RECRYSTALLIZATION

Further Treatment for Maximum Ductility. Although some of the major distortions are eliminated by treatment for recovery, most of the distorted crystalline lattice remains as it was produced by cold work. The elastic limit for the material has been raised close to the ultimate strength, and further deformation will cause fracture failure. Recovery of ductility to permit further change of shape by deformation can be obtained only by elimination of the deformed grains, and this can be accomplished by recrystallization. By this heat-treating process, new, smaller, unstrained grains with fully recovered capacity for plastic flow can be formed by solid-state change in the metal. It is important to note that in the absence of allotropic changes, which will be discussed later, no grain-size changes by heating metal to any temperature below the melting point can be accomplished unless the strained condition of cold-worked metal is present. Recrystallization is the nucleation and growth of new, strain-free crystals from the strained crystals of a cold-worked material.

Recrystallization Temperatures. The phenomenon occurs over a wide temperature range with the length of time required for complete recrystallization inversely related to the temperature and to the degree of strain present. For practical purposes, recrystallization temperatures, such as shown in Table 4-1, are temperatures which will permit complete recrystallization in a time period of approximately 1 hour for metals that have been fully hardened by previous cold work.

TABLE 4-1
Recrystallization Temperatures for Some Common Metals and Alloys

Material	°C	°F
Aluminum (pure)	80	175
Aluminum alloys	316	600
Copper (pure)	120	250
Copper alloys	316	600
Iron (pure)	400	750
Low carbon steel	540	1000
Magnesium (pure)	65	150
Magnesium alloys	232	450
Zinc	10	50
Tin	−4	25
Lead	−4	25

The table shows that zinc, tin, and lead recrystallize at temperatures below room temperature. This means that these metals in the pure state cannot, at ordinary temperatures, maintain a work-hardened condition. The normal use of deformation processes on these materials would be hot working rather than cold working since it would be performed above their recrystallization temperatures. Examination of the table also reveals that contamination of a pure metal with other elements makes it more difficult for recrystallization to occur, and the temperatures must be increased for completion to occur in a reasonable length of time.

Theory of Recrystallization. It is believed that recrystallization takes place by the nucleation of new grains mainly about the high energy points of dislocation in a work-hardened grain. They then appear to grow until they fill the old grain space and eliminate the existing strain by realignment of the atoms into a new crystal lattice. Recrystallization can thus be a grain-refining process as well as a method for recovery of ductility, if it is discontinued as soon as complete recrystallization has taken place.

The new grains formed during recrystallization are likely to take positions with preferred orientations. Directional properties caused by preferred orientation are objectionable for most manufacturing operations. This tendency can be reduced and more random orientation obtained by the addition of small amounts of an alloying element or by recrystallizing before maximum work hardening has been performed.

Recrystallization Seldom Terminal. In a few cases, recrystallization may be used as an end process to leave a product in its most ductile condition or with its best electrical and chemical properties, but more often it is an in-process treatment for ductility improvement or for grain refinement. In many cold deformation processes, such as deep drawing, the ductility of the material may be reduced by cold working to the point where fracture failure is imminent. Ductility may be returned to the material any number of times by repeated recrystallization between steps of the forming operation. In most cases the last forming operation will not be followed by recrystallization, in order that the higher hardness and strength of the cold-worked material may be retained in the product.

Although heating for recovery is a stress-relieving process, recrystallization at a higher temperature is sometimes also called stress relieving. The same process may be referred to as *process annealing,* particularly when performed in conjunction with deformation processes.

GRAIN GROWTH

If a metal is kept heated at or above its recrystallization temperature after the new, unstrained grains have formed, the tendency is for some of the new grains to absorb others and grow to larger size. Large grains are more stable than small grains because of the higher grain-to-boundary-area ratio, which is a lower energy state. If fine grain structure is desired after the recrystallization process, it is necessary to reduce the temperature quickly to prevent subsequent grain growth. This is usually performed by some kind of quench.

Grain-Size Control. During processing, small grain size is not always wanted because large grains usually exhibit greater ductility, better machinability, and require less pressure to be deformed. The final product usually should be of relatively fine structure, though, in order that the material will exhibit its best properties. Grain size for materials that do not go through allotropic phase changes is controlled primarily during the solidification process for cast metals and by recrystallization for wrought (deformation worked) metals. Allotropic metal (existing in more than one crystalline form) grain size can be controlled by a more effective and satisfactory method discussed later in the chapter.

AGE HARDENING

Some metal alloys display a variable solid state solubility of one metal in another with change of temperature. If the solubility increases with increase of temperature above room temperature and if return to the normal room temperature state can be prevented by sudden cooling, the alloy may be susceptible to *age hardening*.

Theory of Age Hardening. Exact explanation of this hardening phenomenon is not available with present knowledge, but from close study it has been theorized that the precipitant from a supersaturated solution first appears as a transition lattice widely dispersed and closely associated with the solid solution lattice. Close association causes lattice distortion with accompanying increase of hardness, much as the distortion by cold working increases hardness. With sufficient time, which decreases with higher temperature, the transition particles combine to form a larger, more widely spaced, and more stable equilibrium precipitant, as in the annealed structure. For hardening purposes, the intermediate phase must be present, and when it disappears because of the complete formation of the final phase, the material is considered to be overaged with loss of the special properties present during the intermediate, or transition, stage.

The need for hardness and strength is often not present at the time of metal solidification. Commercial practice handles age hardening, precipitation hardening, or solution hardening (all names used to describe the same process) as a treatment separate from solidification when there is a need for development of hardness properties or strength properties, or both.

Solution Heat Treatment. The first step is solution treatment (heating) to dissolve a maximum amount of equilibrium precipitant in the solid solution and freeze it in place by sudden cooling to eliminate the necessary time at temperature for precipitation to reoccur. The solution temperature used should be low enough to prevent excessive grain growth but high enough to insure maximum diffusion of the precipitant to saturate the a phase in a minimum amount of time. The time required depends upon the metal alloy and may vary from a few minutes to several hours of soaking at the increased temperature. After saturation of the a phase, the metal is quenched to create the supersaturated solid solution at room temperature. High energy points in the crystal lattice set up by the nonequlibrium situation of supersaturation causes the alloy to be harder than its annealed condition.

Transition Stage by Precipitation. The full hardness, however, is developed during the second stage of treatment when the excess metallic component is partially precipitated from the solid solution. This step is usually referred to as aging and may be natural or artificial. If the surplus material goes into the initial transition stage of precipitation of its own accord at room temperature, full hardness will develop naturally with the passage of time. If an increase of temperature is necessary, as is true with many alloys, to release the unnaturally held metal, this heat-treating step is called artifical aging. Too high an aging temperature or too much time with this stage, or both, causes the precipitant to reach its final equilibrium state in which the hardness and strength properties are low and similar to those of the annealed alloy.

Process Valuable for Aluminum Alloys. One of the greatest uses for precipitation hardening is for improvement of properties of some aluminum alloys. The system can be used for either cast or wrought shapes and can be of particular value in some instances because of the time that is necessary for full hardening to develop. For example, it has been common practice in the aircraft industry to solution treat aluminum rivets and hold them under refrigeration after their quench to retard precipitation. Before precipitation starts, they are relatively ductile and easy to form plastically. In this condition they can be headed to join riveted assemblies and develop their full strength by aging after being upset in place.

ALLOTROPIC CHANGES

Phase Changes. A few metals change lattice structure upon heating and cooling to exist in different forms through various temperature ranges. Such metals are classed as allotropic. Allotropic changes are very similar to the phase changes from liquid to solid, although they occur completely in the solid state with a slower reaction. In addition to a significant change of properties, heat is given up or absorbed as the metal phase change occurs in the solid state but to a much lower degree than in freezing or melting. With some metals special methods are necessary to detect heat changes that accompany the solid-state phase change.

Iron combined with carbon and sometimes small amounts of other elements is by far the most used metal for manufacturing. Iron is an allotropic material that changes upon heating to 912° C (1,674° F) from a body-centered cubic (BCC) lattice to a face-centered cubic (FCC) lattice. A second phase change occurs with further heating to 1,394° C (2541° F), where the lattice structure returns to the body-centered cubic form. The reverse transformation occurs on cooling through the same temperatures. Iron in the temperature range up to 912° C is called alpha iron; from 912° C to 1,394° C, gamma iron; and above 1,394° C to 1,538° C (2,800° F), the melting point, delta iron. Little attention is given to delta iron because the changes that occur in this range have little or no effect in commercial practice of treatment for properties. The changes that take place between alpha and gamma iron at 912° C, however, are extremely important. The most effective change is the difference of carbon solubility in the two phases, which serves as the basis for all heat-treat hardening and most grain-size control for steel.

HEAT TREATMENT OF STEEL

Steel has been treated by heating and cooling methods to vary its properties ever since its discovery, but even today the exact mechanism by which these variations take place cannot be completely explained by fully accepted theories. Most of the treatments have been developed empirically. Various theoretical explanations have been used to describe the mechanism, but it has been only in recent years that the theory has advanced to the point that it is a prime source of new development of commercial heat-treating methods.

NDT and Other Control Methods. Change of properties of steel can be accomplished by cold working, by precipitation hardening, and by allotropic changes. Cold working changes are important in most of the cold deformation processes and, in some cases, may be the only treatment received by the metal. Precipitation hardening is seldom used intentionally, except for stainless steels, although it may be an accidental occurrence with some of the processing treatments. Causing allotropic changes by heat treating procedures is the most effective and most easily accomplished method of varying mechanical properties of steel and therefore is the most frequently used way of obtaining the desired properties.

Heat treating is often defined as intentional heating and cooling for control of properties. Such a definition is perfectly good, but it must be remembered that the effects of temperature changes are no less important when they are caused by unintentional heat transfer during a process such as fusion welding or during a service use in high environmental temperatures such as in a furnace or gas turbine.

Assessment of thermal treatment, whether intentional or not, is often amenable to nondestructive testing techniques that are capable of measuring subtle changes in electrical conductivity. The heat treatment processes described in this chapter produce various physical property changes including electrical conductivity. Both eddy current and thermo-electric methods are capable of indicating changes in electrical conductivity and to some extent can provide absolute measures of electrical conductivity. However, both methods only probe relatively small volumes of the test material essentially at an exposed surface. During heat treatment, exposed surfaces tend to heat and cool at a different rate from the interior. Thus, measurements of surface characteristics do not necessarily characterize the condition of the interior, but in many practical cases can provide adequate information for process control purposes.

APPROXIMATE EQUILIBRIUM HEAT-TREATMENT PROCESSES

Several heat-treating processes place the material in either a complete or an approximate equilibrium energy condition. These processes include austenitizing, annealing, normalizing, and spheroidizing. Except for the first, all are finalized at room temperature, but since austenitizing consists of diffusion of carbon into face-centered cubic iron that exists at a minimum temperature of 727° C (eutectoid composition only, all others higher), stability, or equilibrium, in this state can be maintained only at the higher temperatures. Austenitization is therefore not a final process but only a step in one of several heat-treating procedures. For these approximate equilibrium processes, it is possible to predict the material behavior from the equilibrium phase diagrams.

AUSTENITIZATION

When steel is heated to or above its *critical temperature* (transformation temperature range), the value of which is dependent upon the alloy percentages, and held at temperature for some period of time, carbon unites in solid solution with iron in the gamma or face-centered cubic lattice form. In this phase, as much as 2% carbon can dissolve at the eutectic temperature of 1,148° C at which the widest range of gamma composition exists.

Grain-Size Control. It is important that the austenitization temperatures not be exceeded more than necessary to accomplish the work in a reasonable length of time because grain growth can occur rapidly as the temperature is increased. One of the important features of austenitization is grain refinement that occurs with the formation of the new face-centered cubic lattice. These new small grains are nucleated with the raising of the metal temperature through the austenite range and will remain small if the temperature is not raised too high or maintained too long. With lowering temperature and decomposition of austenite into the room temperature phase, the grain size changes little. Grain sizes are affected only by increasing temperature through this range and not by decreasing temperature. However, because metal grains must be of a certain critical size before they can maintain themselves alone, practically all the grain refinement that is possible can be acquired by one or two austenitization treatments, providing grain growth is not allowed at the higher temperature.

ANNEALING

Objectives of Annealing. The word *anneal* has been used before to describe heat-treating processes for softening and regaining ductility in connection with cold working of material. It has a similar meaning when used in connection with the heat treating of allotropic materials. The purpose of full annealing is to decrease hardness, increase ductility, and sometimes improve machinability of high carbon steels that might otherwise be difficult to cut. The treatment is also used to relieve stresses, refine grain size, and promote uniformity of structure throughout the material.

NORMALIZING

The purpose of *normalizing* is somewhat similar to that of annealing with the exceptions that the steel is not reduced to its softest condition and the pearlite is left rather fine instead of coarse. Pearlite is a crystalline structure with layers of soft, ductile ferrite (iron containing small amounts of dissolved carbon) and hard, brittle cementite (iron carbide which is a mechanical mixture of iron with greater amounts of carbon). Refinement of grain size, relief of internal stresses, and improvement of structural uniformity together with recovery of some ductility provide high toughness qualities in normalized steel. The process is frequently used for improvement of machinability and for stress relief to reduce distortion that might occur with partial machining or aging. An attempt is made during normalizing to dissolve all the cementite to eliminate, as far as possible, the settling of hard, brittle iron carbide in the grain boundaires. The desired decomposition products are small-grained, fine pearlite with a minimum of free ferrite and free cementite.

SPHEROIDIZING

Minimum hardness and maximum ductility of steel can be produced by a process called *spheroidizing*, which causes the iron carbide to form in small spheres or nodules in a ferrite matrix. In order to start with small grains that spheroidize more readily, the process is usually performed on normalized steel. Several variations of processing are used, but all require the holding of the steel near the A_1 temperature (usually slightly below) for a number of hours to allow the iron carbide to form in its more stable and lower energy state of small, rounded globules.

The main need for the process is to improve the machinability quality of high carbon steel and to pretreat hardened steel to help produce greater structural uniformity after quenching. Because of the lengthy treatment time and therefore rather high cost, spheroidizing is not performed nearly as much as annealing or normalizing.

HARDENING OF STEEL

Austenitization — First Step. Most of the heat treatment hardening processes for steel are based on the production of high percentages of martensite. The first step, therefore, is that used for most of the other heat treating processes — treatment to produce austenite.

Fast Cooling — Second Step. The second step involves cooling rapidly in an attempt to avoid pearlite transformation. The cooling rate is determined by the temperature and the ability of the quenching media to carry heat away from the surface of the material being quenched and by the conduction of heat through the material itself. Table 4-2 shows some of the commonly used media and the method of application to remove heat, arranged in order of decreasing cooling ability.

TABLE 4-2
Heat-Treating Quenching

Media	Method
1. Brine	1. Blast
2. Water	2. Violent agitation
3. Light oil	3. Slow agitation
4. Heavy oil	4. Still
5. Air	

Care Necessary in Heating and Cooling. High temperature gradients contribute to high stresses that cause distortion and cracking, so the quench should be only as extreme as is necessary to produce the desired structure. Care must be exercised in quenching that heat is removed uniformly to minimize thermal stresses. For example, a long slender bar should be end-quenched, that is, inserted into the quenching medium vertically so that the entire section is subjected to temperature change at one time. If a shape of this kind were to be quenched in a way that caused one side to drop in temperature before the other, change of dimensions would likely cause high stresses producing plastic flow and permanent distortion.

Cracks created by either heating or quenching can be detected by various NDT surface examinations including eddy current, magnetic particle, and the penetrant methods. For parts subject to fatigure failure, such cracks can be serious.

Distortion and Cracking Minimized by Martempering. Several special types of quench are conducted to minimize quenching stresses and decrease the tendency for distortion and cracking. One of these, is called *martempering* and consists of quenching an austenitized steel in a salt bath at a temperature about that needed for the start of martensite formation. The steel being quenched is held in this bath until it is of uniform temperature but is removed before there is time for the formation of bainite to start. Completion of the cooling in air then causes the same hard martensite that would have formed with quenching from the high temperature, but the high thermal or "quench" stresses that are the primary source of cracks and warping will have been eliminated.

Austempering — a Terminal Step. A similar process performed at a slightly higher temperature is called *austempering*. In this case the steel is held at the bath temperature for a longer period, and the result of the isothermal treatment is the formation of bainite. The bainite structure is not as hard as the martensite that could be formed from the same composition, but in addition to reducing the thermal shock to which the steel would be subjected under normal hardening procedures, it is unnecessary to perform any further treatment to develop good impact resistance in the high hardness range.

TEMPERING

A third step usually required to condition a hardened steel for service is *tempering*, or as it is sometimes referred to, *drawing*. With the exception of austempered steel, which is frequently used in the as-hardened condition, most steels are not serviceable "as quenched". The drastic cooling to produce martensite causes the steel to be very hard and to contain both macroscopic and microscopic internal stresses with the result that the material has little ductility and extreme brittleness. Reduction of these faults is accomplished by reheating the steel to some point below the lower transformation temperature. The structural changes caused by tempering of hardened steel are functions of both time and temperature, with temperature being the most important. It should be emphasized that tempering is not a hardening process, but is, instead, the reverse. A tempered steel is one that has been hardened by heat treatment and then stress relieved, softened, and provided with increased ductility by reheating in the tempering or drawing procedure.

CORROSION

Corrosion Definition. In general, *corrosion* is the deterioration of metals by the chemical action of some surrounding or contracting medium which may be liquid, gas, or some combination of the two. To some degree, corrosion can influence all metals, but its effect varies widely depending upon the combination of metal and corrosive agent.

The term "corrosion" is used to describe action that is normally considered to be detrimental, but the principle is actually used for benefit in some cases. For example, acids and alkalies are used to corrode metal away in the manufacturing process of chemical milling. Also, aluminum alloys are frequently *anodized* to produce an oxide coating that resists further oxidation and, in addition, may serve as an improved surface for paint adhesion.

Corrosion attacks metals by direct chemical action, by electrolysis (electrochemical action), or commonly by a combination of the two. The subject is complex and many persons have devoted their lives to its study. This discussion will summarize some of the known facts concerning the subject in order to develop some understanding of corrosion, its detection, and prevention.

DIRECT CHEMICAL ACTION

Theoretically, all corrosion phenomena are electromechanical because a transfer of electrons takes place but the term *direct chemical action* is used to describe those reactions where coupled anodes and cathodes existing in an electrolyte are not identifiable. The chemical milling mentioned above is direct chemical action.

Another example is pickling of steel, a process in which heated dilute sulpheric acid baths are used to dissolve surface scale without leaving a residue and producing only minor chemical attack on the steel proper. Figure 4-5 illustrates another example of direct chemical action.

Figure 4-5
Direct chemical attack of nitric acid in which a magnesium alloy product is immersed. Direct chemical reaction is usually evident from bubbles formed by gas evolution.

Galvanic Series. Table 4-3 shows a list of metals arranged in order of their decreasing chemical activity in sea water. This is a special arrangement of the electromechanical and the electromotive force series. It

TABLE 4-3
Galvanic series of some metals in sea water

Anodic (Most Corrodible)
Magnesium
Aluminum
Aluminum—Cu Alloy
Zinc
Iron
Steel
Tin
Lead
Nickel
Brass—CuZn
Bronze—CuSn
Copper
Stainless Steel
Silver
Gold
Platinum
Cathodic (Least Corrodible)

should be noted that most of the list is made up of pure metals and indicates their relative resistance to sea water corrosion. If the metals are alloyed or if the corroding medium is different, the arrangement of such a list might change somewhat. In general, a metal high in the series will displace from solution a metal lower in the series.

ELECTROLYTIC (ELECTROCHEMICAL) REACTION

The electrochemical type corrosion also involves chemical change but involves the flow of an electric current between two electrodes, an anode (positive, where electrons leave and negative ions are discharged) and a cathode (negative, where electrons enter and negative ions are formed). An electrical contact must exist in addition to electron flow through the electrolyte to complete the circuit. The system is analogous to a plating system in which the anode supplies the metal to be deposited. The anode eventually is depleted thereby.

Sacrificial Metals. Although there are other factors that influence corrodibility, at least theoretically the metals high in a galvanic series, which are anodic to any metal below them, when connected electrically both by contact and through an electrolyte will dissolve while the cathode is protected. This is the basic use of zinc coatings on steel. The zinc is attacked and sacrificed in order to protect the steel. Protection will continue as long as exposed areas of steel do not grow large enough to develop their own galvanic cells to cause corrosion. The larger the anode area, the better is the protection.

The same principle is used when magnesium rods are hung in hot water heaters to lower corrosion of the tank. Large anodes of magnesium, aluminum, or zinc may be attached to the steel hull of a ship to provide protection below the waterline, as shown in Figure 4-6. Buried steel pipe also may be protected by attaching anodes as shown in Figure 4-7.

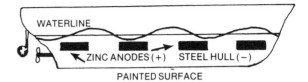

Figure 4-6
A ship's steel hull may be protected by attachment of sacrificial anodic plates to the sides under the water line

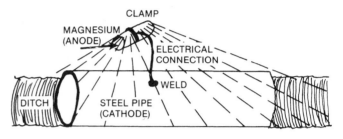

Figure 4-7
Preferential corrosion resulting in protection for buried steel pipe by electrical attachment of anodic material to the steel

Except when sacrificial corrosion protection is planned, it is normally not good practice to design products with contacting metals of radically different galvanic position if there is likelihood of exposure to any corrosive medium. See Figures 4-8 and 4-9 which illustrate a possible lack of good design judgement.

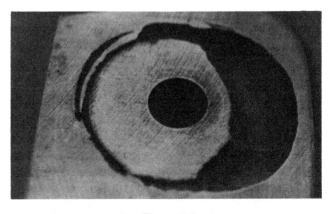

Figure 4-8
Electrochemical corrosion of an aluminum part that was assembled against a steel washer in an environment containing moisture. Being anodic to steel, the aluminum dissolved as shown.

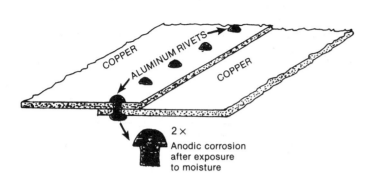

Figure 4-9
Corrosion likely when dissimilar metals are in intimate contact

Galvanic Cells. The system described above causes the flow of electrical current that in turn causes and accelerates corrosion. A type of battery called a galvanic cell can be made of electrodes of two different metals immersed in an electrolyte. A similar result (current flow) is produced when two similar metals are joined, or even a single metal, when contact is made with an electrolyte that is not chemically uniform. This type of cell is known as a concentration cell and is particularly detrimental when the chemical variation of the electrolyte is in its oxygen concentration. Figure 4-10 illustrates the results from this type reaction.

Figure 4-10
Electrochemical corrosion can occur with contacting similar metals when the concentration of the electrolyte varies. Most likely under stagnant conditions.

CORROSION RATE DEPENDENT ON SEVERAL FACTORS

Metal or Metals of a Corrosion System.
1. Position in the electrochemical series.—The higher, the greater tendency for corrosion. In multimetal systems—the farther apart, the greater the electrochemical action.
2. The presence of residual stresses such as shown in Figure 4-11.

Electrolyte Present.
1. Concentration—High concentration usually increases corrosion.

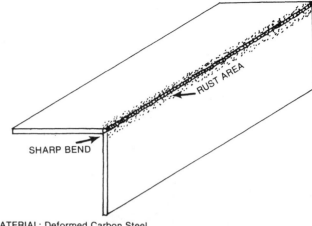

MATERIAL: Deformed Carbon Steel
EXPOSURE: Humid Atmosphere

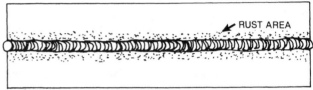

MATERIAL: Butt Welded Carbon Steel
EXPOSURE: Humid Atmosphere

Figure 4-11
Some common examples of corrosion in materials that have localized high stresses

2. Oxygen content—Oxygen particularly harmful in corrosion of iron.
3. Acidity—In general, the higher the acid content, the higher the corrosion rate.
4. Motion—Velocity of a flowing electrolyte may move corrosion products exposing new metal to attack. Movement of electrolyte also may prevent formation of concentration cells, thus reducing corrosion.
5. Temperature—Increase usually accelerates corrosion.
6. Stray electrical currents—Localized currents from leaks, grounds, or eddy currents usually accelerate corrosion.

Atmospheric Corrosion. Moisture is usually blamed for atmospheric corrosion and although moisture may be present, *pure* water has relatively small effect. The combination of moisture with impurities, especially salts of chlorine and sulphur, accelerates atmospheric corrosion greatly.

TYPES OF CORROSION

General Corrosion. The most common type corrosion is that appearing relatively uniformly over the entire surface of the exposed metal. The bluish green color of a copper roof or the dulling of polished aluminum and brass are examples of general corrosion. Some of this type corrosion is self-limiting because the products of early corrosion inhibit further corrosion.

Pitting. Pitting is a localized corrosion by which pits that extend deep into the metal develop. This is a more serious corrosion than the slower general type because the pits may decrease the material strength and also be the nuclei for fatigue failure. With some materials pitting rate may increase with time. Steel which normally rusts uniformly upon exposure to atmosphere may, with sufficient time, develop pits. Figure 4-12 illustrates pitting.

Figure 4-12
Pit type corrosion can be observed in the cylindrical machined surface of this aluminum casting

Intercrystalline Corrosion. A serious type of corrosion is created when the attack is against the grain boundaries. Following the grain boundaries from the metal surface, a crack-like discontinuity develops. Such cracks can cause material failure under static loading by reduction of load supporting cross-section. In the case of dynamic loading, they are likely to be the beginning source of fatigue failure. Because those cracks are seldom visually apparent on the surface, NDT may be called upon for their detection. The sketches of Figure 4-13 illustrate the three main types of corrosion attack.

UNIFORM DISCOLORATION or
LOSS OF POLISH

GENERAL

RANDOM DEEP PITS
SOMETIMES ACCOMPANIED
BY DISCOLORATION

PITTING

CRACK-LIKE DISCONTINUITIES
ALONG GRAIN BOUNDARIES

POLISHED, ETCHED, AND
MAGNIFIED CROSS-SECTION

INTERCRYSTALLINE

Figure 4-13
Principal types of corrosion

Some variations of intercrystalline corrosion are known by the names of *season cracking, stress corrosion* (see Figure 4-14), and *fretting,* all of which are corrosion systems in which corrosion is accelerated by the metal being under load at the same time corrosion is occurring. Season cracking is associated with brass and some other copper-bearing alloys and occurs most frequently when the material has undergone cold workings. Season cracking is much accelerated when the corrosive atmosphere contains ammonia. Fretting is corrosion-assisted wear resulting from small oscillatory movements between mating surfaces under load. Stress corrosion cracking is of major concern because of its effect on a fairly large number of common alloys of various metals used in chemically aggressive environments. In high strength steels and martensitic stainless steels, stress corrosion cracking is usually intergranular; in austenitic stainless steel, usually transgranular. Control of stress corrosion cracking necessitates controlling the four equitial requirements for stress corrosion to occur: a susceptible alloy; an aggressive, conrrosive environment; applied or residual stress; and time. Acoustic emission monitoring techniques have been used in-situ to detect and record the progression of cracking due to stress corrosion.

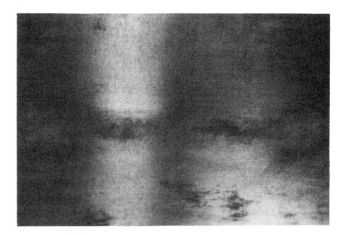

Figure 4-14
Stress-corrosion residue shows on the surface of this magnesium part which has been under constant static load in the presence of a corroding atmosphere

CORROSION PROTECTION

There is no simple answer to preventing serious problems from the attack of corrosion. There is no cure-all because of the variety of metals, possible environments, and corrosive media. The general combat methods include: selection of the most suitable metals, treatment of or controlling the presence of the corroding media, coating the metal with a protective layer, and occasionally stress relieving parts containing high stress areas.

The first is basically a design problem. The second is usually a manufacturing or service problem such as treating coolant water used in a machine tool with a chemical corrosion inhibiter or decerating boiler feed water to remove oxygen. The third is the most common approach and includes: coating with anodic materials to promote preferential corrosion, developing a coating to retard corrosion, and application of a coating to exclude the corrosion medium. The coatings used are metals, chemical compounds, and organic materials and plastics.

Metal Coatings. Coating of metal with another metal can be accomplished by electroplating, dipping in molten metal, metal spraying, cladding by rolling thin layers over the base metal and by heating the product in fine metallic powders.

Chemical Compounds. Most coatings that consist of chemical compounds are made by treating the base metal to change the chemistry of its surface. Anodizing of aluminum is the artificial formation of aluminum oxide to a controlled depth on the surface of an aluminum alloy. Steel can be given a protective coating of iron phosphate by soaking the product in hot solution of manganese phosphate.

Non-metallic Coatings. Paint, enamel, varnishes, greases, plastics, and many other materials are used to coat objects for corrosion protection. Most of these materials are used to exclude the corrosive environment but some contain chemical inhibitors to exert greater control. Some are for only temporary protection such as for a few days or weeks, but others may have a useful life of several years.

In all cases of corrosion protection regardless of the type, suitable preparation and cleaning of the original metal surface is essential. Where control of coating thickness is important, several methods of NDT are available. Eddy current lift-off techniques are most readily applied, but depending upon the type of coating and substrate, beta-backscatter, magnetic field and radioisotopic tagging procedures have been used effectively.

NDT for Corrosion Detection. As apparent from the foregoing discussion of corrosion, its effects are almost always detrimental to the serviceability of critical components, assemblies, and structures. The NDT specialist must understand the effects of the various types of corrosion in order to properly select and direct the nondestructive tests most effective in detecting and assessing the extent of corrosion.

For corroded surfaces that are accessible, penetrant, magnetic particle, and eddy current tests are particulary useful in detecting the effects of corrosion that result in small surface cracks or pits. Very small corrosion cracks have been detected and recorded by magnetic rubber techniques. For corrosion on the inside of pipes, vessels, and assemblies, other tehcniques are applied. Ultrasonic techniques are particularly effective in the detection and accurate measurement of overall thinning that results from corrosion. Radiography is commonly applied to detect corrosion and corrosion thinning in interior and otherwise inaccessible regions of assemblies, insulated components, and the like. Neutron radiography has been used to detect interior corrosion by virtue of the corrosion products having large neutron cross-sections as well as actually imaging corrosion in exceptionally dense materials like lead and uranium. Acousic emission monitoring has been used to monitor the initiation and growth of stress corrosion and hydrogen embrittlement cracks.

Visual means are also important in the detection of corrosion. Both corrosion discontinuities and corrosion products leave telltale signs by virtue of visible changes in texture, coloration, topography, and geometry. Some corrosion products fluoresce when illuminated by ultraviolet light. Further study of corrosion sites and corrosion products by spectrographic analyses can reveal otherwise elusive evidence as to the cause of corrosion.

Ferrous Metals 5

CHOOSING METALS AND ALLOYS

In Chapter 4 metals were discussed primarily on the basis of their atomic configurations. While it is true that this basis gives a more precise definition in the chemist's or physicist's terms, of greater practical interest in manufacturing are the metallic properties of relatively high hardness and strength, ability to undergo considerable plastic flow, high density, durability, rigidity, luster. A distinction is sometimes made between the word *metal*, meaning a pure chemical element, and the word *alloy*, meaning a combination of materials, the predominant one of which is a metal. The term *metal* in this text will be taken to mean any metallic material, whether pure or alloyed.

Availability of Ores. Among all the possible reasons for the choice and use of a material, one of very prime importance is availability. Table 5-1 shows the composition of the earth's crust. Of the first twelve elements in occurrence, aluminum, iron, magnesium, and titanium are used as the base metals of alloy systems. For the other metals, although the total tonnage in the earth's crust may be considerable, the potential use is much more restricted. Some of them, such as copper, are found in relatively pure deposits but frequently in remote locations, and the total use is dependent on relatively few of these rich deposits. Most other metals are recovered only in relatively small quantities, either as byproducts of the recovery of the more predominant metals or as products of low-yield ores after extensive mining and concentration in which many tons of material must be handled for each pound of metal recovered. The United States has only marginal deposits of antimony, chromium, cobalt, manganese, and nickel and imports the major quantity of these metals. It is almost totally dependent on imports for its supply of mercury, tungsten, and tin. The location and the availability of these materials have a marked influence on both the risk and cost of choosing these materials for large-use applications.

TABLE 5-1
Elements in the Earth's Crust

Element	Percent	Element	Percent
Oxygen	46.71	Magnesium	2.08
Silicon	27.69	Titanium	0.62
Aluminum	8.07	Hydrogen	0.14
Iron	5.05	Phosphorus	0.13
Calcium	3.65	Carbon	0.094
Sodium	2.75	Others	0.436
Potassium	2.58		

Base Metals. Approximately seventy of the elements may be classed as metals, and of these, about forty are of commercial importance. Historically, copper, lead, tin, and iron are metals of antiquity because they are either found free in nature or their ores are relatively easy to reduce. These four metals together with aluminum, magnesium, zinc, nickel, and titanium are presently the most important metals for use as base metals for structural alloy systems. Most other commercially important metals either are metals used primarily as alloying metals or noble metals, such as gold, silver, or platinum, that are important only for special uses or because of their rarity.

Material Choice Affected by Process. The method of manufacture will frequently affect the alloy type chosen even after the base metal has been chosen. Although nearly all metals are cast at some time during their manufacture, those that are cast to approximate finished shape and finished without deformation are specifically referred to as *casting alloys*. When the metal is fabricated by deformation processes, an alloy designed to have good ductility is specified and referred to as a *wrought alloy*. Some alloys can be either wrought or cast, most wrought alloys can be cast, but many casting alloys have insufficient ductility for even simple deformation processing.

Final Choice Dependent on Many Factors. The choice of a material is usually a stepwise process. Sales requirements, raw material costs, equipment availability, or specific product requirements will frequently narrow the choice between the fields of metals and plastics. With the choice of either metals or plastics, some may be eliminated on the basis of properties, although a considerable number of plastics or metal alloys will still satisfy the functional requirements for the great majority of products. The life to be expected from the product may also eliminate some materials from consideration. Finally, however, the choice usually becomes one based on costs. From the various materials that would produce a functionally acceptable product with sufficient life and from the various processing methods that are available to a manufacturer, the best combination must be found. Obviously, many combinations will be rather quickly eliminated, but of those remaining, costs of some may not be entirely predictable without actual experience in producing the product. Consequently, the first choice is not always the final choice, and for this reason, as well as for reasons of sales appeal and product redesign, materials and processes frequently are changed on a trial and error basis.

Importance of Ferrous Materials. The role that ferrous materials play in the economy is evident from annual production figures. Approximately 100 million tons of ferrous products are made each year in the United States. For all nonferrous metals, the total is about 10 million tons per year. Even though much of the steel tonnage goes into heavy products such as rails and structural steel shapes that require little secondary work, ferrous metals are still the predominant materials of manufacturing. The wide variety of ferrous products is based largely on the economy of producing them; an attempt will be made to discuss ferrous metals in the economic order of their production in the section to follow. Generally, as better properties are required, more costly processes are necessary.

FERROUS RAW MATERIALS

Ore Reduction. Both iron and steel have their start in the blast furnace. Although other methods for reduction have been proposed and will likely be developed, the tremendous investment in equipment and trained personnel that would be required for the replacement of present facilities almost insures that the blast furnace method will remain for some time.

This device is a tall, columnar structure into which is fed, through a top opening, a mixture of iron ore (oxides of iron — Fe_3O_3, hematite, or Fe_3O_4, magnetite), coke, and limestone. A blast of hot air is supplied through the mixture from near the bottom to provide oxygen for combustion of the coke. Temperatures in the neighborhood of 3000° F are developed in the melting zone. The iron ore is reduced by chemical reactions with carbon monoxide gases and by high temperature contact directly with the carbon in the coke as well as with other impurity elements in the mixture. Near the bottom of the furnace, the iron and the slag, which is made up of other metallic oxides combined with limestone, melt and accumulate in a well; the lighter slag floats on top of the melted iron. The molten iron and slag are tapped off periodically through separate holes. The slag is disposed of, either as trash or for byproduct use, and the iron is run into open molds to solidify as *pigs*, unless it is to be further processed immediately. In large installations, the molten iron is frequently transported in large ladles to other equipment for carbon reduction in the manufacture of steel.

Pig Iron. The product of the blast furnace, whether liquid or solid, is called *pig iron*. The distinction between the terms *pig* and *pig iron* should be noted. The term *pig* refers to a crude casting, convenient for transportation, storage, and remelting of any metal: the term *pig iron* refers to the composition of the metal tapped from the blast furnace, whether in liquid or solid state. Although this composition varies with ore, coke, blast furnace conditions, and other factors, the blast furnace is controllable only within broad limits. Pig iron as a natural result of the conditions within the furnace always contains 3% to 4% of carbon and smaller amounts of silicon, sulfur, phosphorus, manganese, and other elements.

Pig Iron Requires Further Processing. In the solid state, pig iron is weak, is too hard to be machined, and has practically no ductility to permit deformation work. It must therefore be treated to improve some of its properties by one of the methods shown in Figure 5-1. The simplest of these treatments are those shown on the left of Figure 5-1; the treatments involve remelting with only moderate control of composition, in particular with no attempt to remove the carbon.

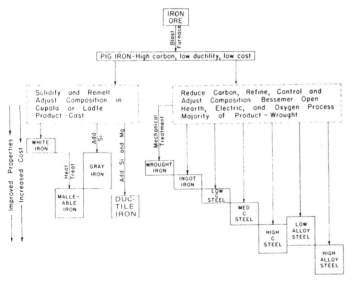

Figure 5-1
General relationship of ferrous materials

CAST IRONS

These simplest ferrous materials are produced by causing the molten metal to solidify into approximate final product form. The result is known as a casting. The processes of making castings is discussed in Chapter 8. Some of the relationships between common cast irons are shown in Table 5-2.

STEEL

One of the largest and most influential manufacturing operations today is the steel industry, which makes some finished products but is primarily concerned with the making of raw material for further processing. The annual production of more than 100 million tons exceeds by far the total production of all other metals and plastics combined.

Comparison of Steel with Cast Iron. Pound for pound, castings of cast iron are cheaper than those of steel, and for those products that can be made with suitable shapes and strengths as castings, the cost of the finished product often will be lower in this form. However, all cast irons, because of their high carbon content, are subject to the definite processing limitations of casting. Thin sections, good finishes, and dimensional control are obtained at reasonable cost only by deformation processing instead of casting. Deformation can be performed only on materials having relatively high ductility. For ferrous materials, this requires reduction of carbon from the cast iron range to the extent that a material with an entirely new set of properties is produced.

All cast irons are essentially pig iron with, at most, only minor modifications of composition. The essential component of pig iron in addition to the iron is 3% to 4% carbon. When this carbon content is reduced to less than 2%, the resulting new material is called *steel*.

WROUGHT IRON

Prior to the introduction of currently used methods for making steel, a method of reducing the carbon content of pig iron had been used since before 1600. The product, although called *wrought iron*, was actually the first low carbon steel to be manufactured in quantity.

Early Furnace Limitations. In the early manufacture of wrought iron, molten pig iron was subjected to oxidizing agents, normally air and iron oxide, and the silicon and carbon content of the melt was reduced. The furnaces used were incapable of maintaining the iron at temperatures greater than about 1480° C (2700° F). Reference to the iron-carbon equilibrium diagram will show that at this temperature pig iron would be well above the liquidus line. However, as the carbon content was reduced, at constant temperature, the iron began to solidify; consequently, to keep the reaction proceeding within the melt, it was necessary to stir or *puddle* the material in the furnace.

Wrought Iron Contains Slag. Because this material included slag, which floated on top as long as the metal was liquid, the slag was mixed with the purified iron. The resulting product was withdrawn from the furnace as a pasty ball on the end of the stirring rod

TABLE 5-2
Common cast irons

Type Iron	How Produced	Characteristics	Relative Cost
White	Rapid cooling Low C + Si	Hard, brittle Unmachinable	1
Malleable	Heat treated White iron	T.S. 3.5-8 × 10^8 Pa (50—120 ksi) Good malleability and ductility	4
Ductile	Ladle addition	T.S. 4-10 × 10^8 Pa (60—150 ksi) Similar to malleable	3
Gray	Slow cooling High C + Si	T.S. 1.4-4.1 × 10^8 Pa (20—60 ksi) Good machinability Brittle	2
Chilled	Fast surface chill	Hard surface (white iron) Soft core (gray iron)	3

and, while low in carbon and silicon, contained from 3% to 4% slag, mostly SiO_2. These balls were then deformation processed by repeated rolling, cutting, stacking, and rerolling in the same direction. The resulting product consisted of relatively pure iron with many very fine slag stringers running in the direction of rolling.

Although cheaper methods have been developed for reducing the carbon from pig iron without incorporating the slag in the product, a demand for wrought iron continues, based primarily on its reputation for corrosion and fatigue resistance. It is presently manufactured by pouring molten refined iron into separately manufactured slag with subsequent rolling.

Properties of Wrought Iron. Wrought iron has a tensile strength of about 350 MPa (50,000 psi) and good ductility, although the material is quite *anisotropic* (properties vary with orientation or direction of testing) because of the slag stringers. Its principal use is for the manufacture of welded pipe.

While *wrought iron* originally referred to this product or to its composition, the term has frequently been extended to refer to any *worked* low carbon steel product, particularly a product shaped or worked by hand, such as ornamental iron railings and grillwork.

STEEL MAKING

Early Steel. The oldest known method of making higher carbon steel consisted of reheating wrought iron and powdered charcoal together in the *cementation* process. According to the iron-carbon equilibrium diagram, at 1148° C (2098° F) carbon is soluble in iron up to 2%. At this temperature the carbon slowly diffused into the solid material; the process required a total cycle time, including heating, of about 2 weeks. Much of the slag in the wrought iron migrated to the surface and formed surface blisters, which resulted in the term *blister* steel. Even after this lengthy treatment, the carbon was not uniformly dispersed throughout the material, and multiple cutting and rerolling procedures were required to produce a high quality product.

Crucible Steel. Further reduction of the slag, greater uniformity of the carbon, and closer control were later achieved by a secondary operation known as the *crucible* process. Bars made by the cementation process were remelted in a clay or graphite crucible in which the slag floated to the surface. This crucible process produced steel of very high quality, and modifications of the method are still used today, but it was made possible only by furnace developments that permitted higher temperatures to be achieved than were needed in the manufacture of wrought iron.

Open-Hearth Steel. Both the modern open-hearth furnace and the Bessemer converter were developed in the 1850s. These two developments greatly increased the speed with which pig iron could be refined. The modern era of industry can be tied to these developments that led to the production of large quantities of high quality, low-cost steel.

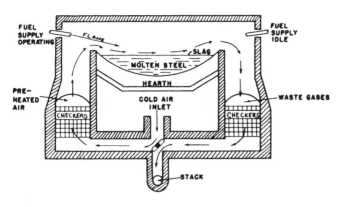

Figure 5-2
Cross-section of open-hearth furnace

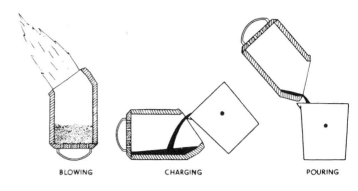

Figure 5-3
Bessemer converter

Figure 5-2 shows the construction of an open-hearth furnace as was used for the majority of steel produced until recently in the United States. Various proportions of pig iron (either solid or molten), steel scrap, limestone for flux, and iron ore are charged on the hearth of the furnace. The principal reducing action takes place between the iron ore and the carbon of the pig iron, the final carbon content of the steel being controllable by the proper proportions of the charged materials. The principal difference between this furnace and that used previously in the manufacture of wrought iron lies in the preheating of the entering combustion air. In the open-hearth furnace for steel making, the air enters through a brick checkerwork that has been previously heated by the exhausting flue gases. Two similar checkerworks are used, one for the exhaust side and one for the entering air side of the furnace. After a relatively short period of operation in this manner, the airflow through the checkerworks is reversed. Preheating of the air permits higher temperatures to be developed in the furnace, and the bath of metal may be kept molten as the carbon content is reduced.

Bessemer Steel. The Bessemer converter is shown in Figure 5-3. The charge consists of molten pig iron. Steel scrap may be added to help control the temperature. After charging in the horizontal position, the air blast is turned on through the *tuyeres* and the converter turned upright so that the air bubbles through the melt, oxidizing and burning out first silicon, then carbon. The process can be used to reduce the carbon content to about 0.05%. Although less expensive to operate than the *basic*-lined open-hearth furnace, the inability of the *acid*-lined Bessemer converter to reduce the phosphorus content of the metal has restricted its use to the production of only about 5% of the steel made in the United States. Some steel is produced by initial refining in the Bessemer converter followed by further refining in the open-hearth furnace.

Electric Furnace Steel. Electric furnace steel is produced in a variation of the older crucible process with the furnace heated by electric arc or induction. The atmosphere can be well controlled in the electric furnace, and careful control of composition can be maintained. Steel of the highest quality is produced by this method.

Basic Oxygen Steel. A steel making process known as the basic oxygen process was developed in Switzerland and Austria after World War II and first used in 1952. By 1957 the method was producing 1% of the world production. In 1966 the growth of use was to 25% and currently more than 50% of the world's steel is made by the basic oxygen process.

The Basic Oxygen Process. There are a number of variations in the equipment and methods for making basic oxygen steel. Fundamentally they all operate much as follows:

a. Scrap as great as 30% of the heat is charged into the refining vessel, as shown schematically in Figure 5-4.
b. Molten pig iron is charged on top of the scrap.
c. The lance is positioned, and a high velocity jet of oxygen is blown on top of the molten mixture for about 20 minutes. During this period, lime and various fluxes are added as aids for control of the final composition.
d. The metal is then sampled, and, if it meets specifications, poured through the tap hole into a ladle by tilting the vessel.
e. Finally, the vessel is inverted to empty the slag and then is ready for reuse. With careful use, the vessel lining may last for as many as 400 heats.

The total time for producing a heat by this method is 30 to 45 minutes. This compares very favorably with the 4 to 6 hours necessary for the open-hearth methods using oxygen.

Basic Oxygen Process Provides a Number of Advantages. Steel made by this method can start from any grade of pig iron. The finish quality is similar to that

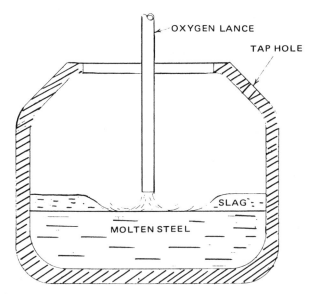

Figure 5-4
Basic oxygen furnace

amount of carbon, the properties approach those of pure iron with maximum ductility and minimum strength. Maximum ductility is desirable from the standpoint of ease in deformation processing and service use. Minimum strength is desirable for deformation processing. However, higher strengths than that obtainable with this low carbon are desirable from the standpoint of product design. The most practical means of increasing the strength is by the addition or retention of some carbon. However, it should be fully understood that any increase of strength over that of pure iron can be obtained only at the expense of some loss of ductility, and the final choice is always a compromise of some degree. Figure 5-5 shows typical ferrous material applications in relation to carbon content. Because of the difficulty of composition control or the additional operation of increasing carbon content, the cost of higher carbon, higher strength steel is greater than that of low carbon.

Plain Carbon Steels Most Used. Because of their low cost, the majority of steels used are plain carbon steels. These consist of iron combined with carbon

made in open-hearth furnaces. Scrap is usable in large quantities so that the process becomes the cheapest current method for remelting and reusing scrap.

The largest size unit presently available is slightly greater than 300 tons. A 300-ton unit can produce 3 million tons of steel per year.

Basic Oxygen Process Limited by Huge Investment Needs. The growth of the basic oxygen process has been extremely fast as industrial processes go but would probably have been even faster except for the large investments required. The immense quantities of oxygen and its use demand much special equipment. In such a conversion to a facility including a rolling mill, one steel manufacturer invested over $600 million.

Practically All Steel Made Today by Use of Oxygen. The development of oxygen-making facilities and the reduction of cost of the gas has changed nearly all steel making. Even when the complete basic oxygen process is not used, oxygen is used to speed steel making. Both open-hearth and Bessemer converters are likely to be supplied with oxygen to speed combustion and refining. An open-hearth furnace fitted with oxygen lances can approximately double production with less than one-half the fuel of earlier methods, without use of pure oxygen. The making of Bessemer steel is speeded by use of oxygen combined with air but also is improved in composition, mainly by reduction of nitrogen impurities left in the steel. Little Bessemer steel is made in the United States, however.

PLAIN CARBON STEEL

Any steel-making process is capable of producing a product that has 0.05% or less carbon. With this small

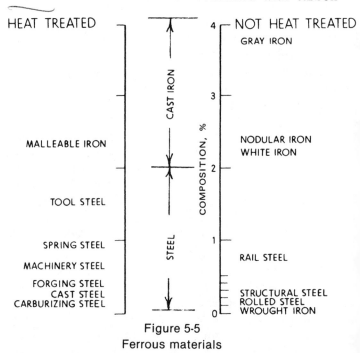

Figure 5-5
Ferrous materials

concentrated in three ranges classed as low carbon, medium carbon, and high carbon. With the exception of manganese used to control sulphur, other elements are present only in small enough quantities to be considered as impurities, though in some cases they may have minor effect on properties of the material.

Low Carbon. Steels with approximately 6 to 25 points *of carbon* (0.06% to 0.25%) are rated as low carbon steels and are rarely hardened by heat treatment because the low carbon content permits so little formation of hard martensite that the process is relatively ineffective. Enormous tonnages of these low carbon steels are processed in such structural shapes as sheet, strip, rod, plate, pipe, and wire. A large portion of the material is cold worked in its final processing to improve its hardness, strength, and surface-finish qualities. The grades containing 20 points or less of carbon are susceptible to considerable plastic flow and are frequently used as deep-drawn products or may be used as a ductile core for casehardened material. The low plain carbon steels are readily brazed, welded, and forged.

Medium Carbon. The medium carbon steels (0.25% to 0.5%) contain sufficient carbon that they may be heat treated for desirable strength, hardness, machinability, or other properties. The hardness of plain carbon steels in this range cannot be increased sufficiently for the material to serve satisfactorily as cutting tools, but the load-carrying capacity of the steels can be raised considerably, while still retaining sufficient ductility for good toughness. The majority of the steel is furnished in the hot-rolled condition and is often machined for final finishing. It can be welded, but is more difficult to join by this method than the low carbon steel because of structural changes caused by welding heat in localized areas.

High Carbon. High carbon steel contains from 50 to 160 points of carbon (0.5% to 1.6%). This group of steels is classed as tool and die steel, in which hardness is the principal property desired. Because of the fast reaction time and resulting low hardenability, plain carbon steels nearly always must be water-quenched. Even with this drastic treatment and its associated danger of distortion or cracking, it is seldom possible to develop fully hardened structure in material more than about 1 inch in thickness. In practice the ductility of heat-treat-hardened plain carbon steel is low compared to that of alloy steels with the same strength, but, even so, carbon steel is frequently used because of its lower cost.

ALLOY STEELS

Although plain carbon steels work well for many uses and are the cheapest steels and therefore the most used, they cannot completely fulfill the requirements for some work. Individual or groups of properties can be improved by addition of various elements in the form of alloys. Even plain carbon steels are alloys of at least iron, carbon, and manganese, but the term *alloy steel* refers to steels containing elements other than these in controlled quantities greater than impurity concentration or, in the case of manganese, greater than 1.5%.

Composition and Structure Affect Properties. Table 5-3 shows the general effects of the more commonly used elements on some properties of steels. Some effects noted in the chart are independent, but most are based on the influence the element has on the action of carbon. The hardness and the strength of any steel, alloy or otherwise, depend primarily on the amount and the form of the iron carbide or other metal carbides present. Even in unhardened steel, carbon produces an increase in hardness and strength with a consequent loss of ductility. The improvement in machinability and the loss in weldability are based on this loss of ductility.

Alloys Affect Hardenability. Interest in hardenability is indirect. Hardenability itself has been discussed earlier and is usually thought of most in connection with depth-hardening ability in a full hardening operation. However, with the isothermal transformation curves shifted to the right, the properties of a material can be materially changed even when not fully hardened. After hot-rolling or forging operations, the material usually air cools. Any alloy generally shifts the transformation curves to the right, which with air cooling results in finer pearlite than would be formed in a plain carbon steel. This finer pearlite has higher hardness and strength, which has an effect on machinability and may lower ductility.

Weldability. The generally bad influence of alloys on weldability is a further reflection of the influence on hardenability. With alloys present during the rapid cooling taking place in the welding area, hard, nonductile structures are formed in the steel and frequently lead to cracking and distortion.

Grain Size and Toughness. Nickel in particular has a very beneficial effect by retarding grain growth in the austenite range. As with hardenability, it is the secondary effects of grain refinement that are noted in properties. A finer grain structure may actually have less hardenability, but it has its most pronounced effect on toughness; for two steels with equivalent hardness and strength, the one with finer grain will have better ductility, which is reflected in the chart as improved toughness. This improved toughness, however, may be detrimental to machinability.

Corrosion Resistance. Most pure metals have relatively good corrosion resistance, which is generally lowered by impurities or small amounts of intentional alloys. In steel, carbon in particular lowers the corrosion resistance very seriously. In small percentages, copper and phosphorus are beneficial in reducing corrosion. Nickel becomes effective in percentages of about 5%, and chromium is extremely effective in percentages greater than 10%, which leads to a separate class of alloy steels called stainless steels. Many tool steels, while not designed for the purpose, are in effect stainless steels because of the high percentage of chromium present.

LOW ALLOY STRUCTURAL STEELS

Certain low alloy steels sold under various trade names have been developed to provide a low cost

TABLE 5-3
Effect of Some Alloying Elements on Properties of Steel

	Low Carbon 0.1%-0.2%	Med Carbon 0.2%-0.6%	Manganese 2.0%	Phosphorus 0.15%	Sulfur 0.3%	Silicon 2.0%	Chromium 1.1%	Nickel 5.0%	Molybdenum 0.75%	Vanadium 0.25%	Copper 1.1%	Aluminum 0.1%	Boron 0.003%
Hardenability	N	G	VG	G	B	G	VG	VG	VG	G	N	N	VG
Strength	G	VG	G	G	B	VG	G	G	G	N	N	G	G
Toughness	B	VB	G	VB	VB	B	VB	VG	G	G	N	G	?
Wear resistance	N	VG	VG	N	N	G	VG	G	VG	G	N	N	G
Machinability annealed	G	G	B	G	VG	B	B	VB	B	N	B	N	?
Weldability	B	VB	VB	VB	B	B	VB	VB	VB	G	B	N	VB
Corrosion resistance	B	VB	N	VG	VB	G	N	VG	G	N	VG	G	?

Very good	VG
Good	G
Little or none	N
Bad	B
Very bad	VB

structural material with higher yield strength than plain carbon steel. The addition of small amounts of some alloying elements can raise the yield strength of hot-rolled sections without heat treatment to 30% to 40% greater than that of plain carbon steels. Designing to higher working stresses may reduce the required section size by 25% to 30% at an increased cost of 15% to 50%, depending upon the amount and the kind of alloy.

The low alloy structural steels are sold almost entirely in the form of hot-rolled structural shapes. These materials have good weldability, ductility, better impact strength than that of plain carbon steel, and good corrosion resistance, particularly to atmospheric exposure. Many building codes are based on the more conservative use of plain carbon steels, and the use of alloy structural steel often has no economic advantage in these cases.

LOW ALLOY AISI STEELS

Improved Properties at Higher Cost. The low alloy American Iron and Steel Institute (AISI) steels are alloyed primarily for improved hardenability. They are more costly than plain carbon steels, and their use can generally be justified only when needed in the heat-treat-hardened and tempered condition. Compared to plain carbon steels, they can have 30% to 40% higher yield strength and 10% to 20% higher tensile strength. At equivalent tensile strengths and hardnesses, they can have 30% to 40% higher reduction of area and approximately twice the impact strength.

Usually Heat Treated. The low alloy AISI steels are those containing less than approximately 8% total alloying elements, although most commercially important steels contain less than 5%. The carbon content may vary from very low to very high, but for most steels it is in the medium range that effective heat treatment may be employed for property improvement at minimum costs. The steels are used widely in automobile, machine tool, and aircraft construction, especially for the manufacture of moving parts that are subject to high stress and wear.

STAINLESS STEELS

Tonnage-wise, the most important of the higher alloy steels are a group of high chromium steels with extremely high corrosion and chemical resistance. Most of these steels have much better mechanical properties at high temperatures. This group was first called *stainless steel*. With the emphasis on high temperature use, they are frequently referred to as heat and corrosion-resistant steels.

Martensitic Stainless Steel. With lower amounts of chromium or with silicon or aluminum added to some of the higher chromium steels, the material responds to heat treatment much as any low alloy steel. The gamma-to-alpha transformation in iron occurs normally, and the steel may be hardened by heat treatment similar to that used on plain carbon or low alloy steels. Steels of this class are called *martensitic*, and the most used ones have 4% to 6% chromium.

Ferritic Stainless Steel. With larger amounts of chromium, as great as 30% or more, the austenite is suppressed, and the steel loses its ability to be hardened by normal steel heat-treating procedures. Steels of this type are called *ferritic* and are particularly useful when high corrosion resistance is necessary in cold-worked products.

Austenitic Stainless Steel. With high chromium and the addition of 8% or more of nickel or combinations of nickel and manganese, the ferrite is suppressed. These steels, the most typical of which contains 18% chromium and 8% nickel, are referred to as *austenitic* stainless steels. They are not hardenable by normal steel heat-treating procedures, but the addition of small amounts of other elements makes some of them hardenable by a solution precipiation reaction.

Composition and Structure Critical for Corrosion Resistance. In any stainless steel, serious loss of corrosion resistance can occur if large amounts of chromium carbide form. Consequently, the ferritic and austenitic grades are generally made with low amounts of carbon and even then may need special heat treatments or the addition of *stabilizing* elements such as molybdenum or titanium to prevent chromium carbide formation. With the martensitic grades in which the hardness and strength depend on the carbon, the steels must be fully hardened with the carbon in a martenistic structure for maximum corrosion resistance.

The austenitic steels are the most expensive but possess the best impact properties at low temperatures, the highest strength and corrosion resistance at elevated temperature, and generally have the best appearance. They are used for heat exchangers, refining and chemical processing equipment, gas turbines, and other equipment exposed to severe corrosive conditions. The austenitic steels are paramagnetic (practically unaffected by magnetic flux). This fact precludes the use of magnetic particle testing. In the as-cast state, and in welds, austenitic stainless steel is quite coarse-grained. In ultrasonic testing of this material, high levels of noise and attenuation serve to limit the effectiveness of the test.

Both the ferritic and martensitic stainless steels are magnetic. Most are not as corrosion resistant at high temperatures as the austenitic type but offer good resistance at normal temperatures. They are used for such products as cutlery, surgical instruments, automobile trim, ball bearings, and kitchen equipment.

Fabrication Difficult. The stainless steels are more difficult to machine and weld than most other ferrous materials. In no case can stainless steels be classed as the easiest to work, but they can be processed by all of the normal procedures, including casting, rolling, forging, and pressworking.

TOOL AND DIE STEELS

The greatest tonnage of tools (other than cutting tools, which are discussed in Chapter 18) and dies are made from plain carbon or low alloy steels. This is true only because of the low cost of these materials as their use has a number of disadvantages. They have low hardenability, low ductility associated with high hardness, and do not hold their hardness well at elevated temperature.

Manganese Steels. Manganese tool and die steels are oil hardening and have a reduced tendency to deform or crack during heat treatment. They contain from 85 to 100 points of carbon, 1.5% to 1.75% of manganese to improve hardenability, and small amounts of chromium, vanadium, and molybdenum to improve hardness and toughness qualities.

Chromium Steels. High chromium tool and die steels are usually quenched in oil for hardening, but some have sufficient hardenability to develop hardness with an air quench. One group of the high chromium steels, called high speed steel, has substantial additions of tungsten, vanadium, and sometimes cobalt to improve the hardness in the red heat range.

CAST STEELS

Quantity Relatively Small. Compared to the tonnage of cast iron and wrought steel produced, the quantity of cast steel is small. The high temperatures necessary make melting and handling more difficult than for cast iron and also create problems in producing sound, high quality castings. The mechanical properties of cast steel tend to be poorer than those of the same material in wrought form, but certain shape and size relationships, together with property requirements that can be supplied only by steel, may favor the manufacture of a product as a steel casting. Steel castings may be produced with greater ductility than even malleable iron.

Cast Steel Is Isotropic. The principal advantages of steel as a structural material, mainly the ability to control properties by composition and heat treatment, apply for both the wrought and the cast material. One advantage of cast steel over its wrought counterpart is its lack of directional properties. Wrought steel and other materials tend to develop strength in the direction of working when deformed by plastic flow, that is, become anisotropic. At the same time, they become weaker and more brittle in the perpendicular directions. Steel that is cast to shape loses the opportunity for gain in properties by plastic work but, by the same token, is not adversely affected by weakness in some directions.

Wide Variety of Composition. As far as composition is concerned, no real differences exists between wrought and cast steel. It was pointed out earlier that steel is a combination of mostly iron with carbon in amounts from just above that soluble at room temperature (0.008%) to as high as 2%, the maximum soluble in austenite at the eutectic temperature. Other elements may also be part of the composition in quantities small enough to be negligible or sufficiently large to influence the heat treating of the alloy or even exert effects of their own, as in wrought alloy steels. The carbon content can be in any of the three ranges, low, medium, or high, but the majority of steel castings are produced in the medium carbon range because nearly all are heat treated to develop good mechanical properties.

MATERIAL IDENTIFICATION SYSTEMS

Variety of Metallic Materials Necessitates Specification Codes. During earlier times in our industrial development, there was less need for material identification systems. A manufacturer generally had complete charge of the entire operation from raw material to finished product. In any event, there were relatively few materials from which to choose. More recently, specialization has led to more division of the manufacturing procedure. Fabricators seldom produce their own raw materials, and the number of material choices has grown tremendously and continues to grow yearly. Reliable and universally accepted systems of material specification are essential to permit designers to specify and fabricators to purchase materials and be assured of composition and properties.

The first group of materials for which standardization was needed was ferrous materials. The automotive industry set up the first recognized standards, but with broader use and more classes of steels, the present most universally recognized standards are those of the AISI.

AISI Numbers for Plain and Low Alloy Steels. The number of possible combinations of iron, carbon, and alloying elements is without limit. Some of these, for example, the low alloy high strength structural steels, are not covered by any standard specification system, or designation. However, the majority of commonly used steels in the plain carbon and low alloy categories can be described by a standardized code system consisting of a letter denoting the process by which the steel was manufactured, followed by four, or in a few cases, five digits. The first two digits refer to the quantity and kind of principal alloying element or elements. The last two digits, or three in the case of some high carbon steels, refer to the carbon content in hundredths of a percent. At one time, the process used in steel making affected the properties of the finished product enough that it was important to know how it was made. Letter prefixes as follows were used for this purpose.

- B — Acid Bessemer carbon steel
- C — Basic open-hearth steel
- D — Acid open-hearth steel
- E — Electric furnace alloy steel

With the advent of basic oxygen steel, however, the letter prefix is falling into disuse. The control exhibited in the basic oxygen process produces steel of similar quality to that from the open-hearth method.

Table 5-4 shows the average alloy content associated with some of the most frequently used classes of steels. The exact specified quantity varies with the carbon content of each steel, and even steels with exactly the same number throughout will vary slightly from heat to heat because of necessary manufacturing tolerances. Exact composition can therefore be determined only from chemical analysis of individual heats.

TABLE 5-4
AISI Basic Classification Numbers

AISI No.	Average Percent Alloy Content
10XX	None
11XX	0.08-0.33 S
13XX	1.8-2.0 Mn
23XX	3.5 Ni
31XX	0.7-0.8 Cr 1.3 Ni
41XX	0.5-1.0 Cr, 0.2-0.3 Mo
43XX	0.5-0.8 Cr , 1.8 Ni, 0.3 Mo
51XX	0.8-1.1 Cr
61XX	0.8-1.0 Cr, 0.1-0.2 V
86XX	0.6 Ni, 0.5-0.7 Cr, 1.2 Mo
87XX	0.6 Ni, 0.5 Cr, 0.3 Mo

TABLE 5-5
Some Stainless Steels and Properties

Material	Composition Ni	Composition Cr	Composition Other	Ten St 1000 psi (6.9 × 10^6 Pa)	Percent Elong (2 in.)	Characteristics and Uses
302	9	18				Austenitic — Work harden only. Excellent corrosion resistance to atmosphere and foods. Machinability fair. Welding not recommended. General purpose. Kitchen and chemical applications.
Annealed				85	60	
430		16	C0.12			Ferritic — Work harden only. Excellent corrosion resistance to weather and water exposure and most chemicals. Machinability fair. General purpose. Kitchen and chemical equipment. Automobile trim.
Annealed				75	30	
Cold worked				90	15	
420		13	C0.15			Martensitic — Heat treatable. Good corrosion resistance to weather and water exposure. Machinability fair. Cutlery, surgical instruments, ball bearings.
Annealed				95	25	
Hardened and tempered				230	8	
17-4PH	4	17	Cu 4			Age hardening — Good corrosion resistance. Maintains strength at elevated temperature. Machinability poor. Air frame skin and structure.
Room temp				195	13	
1200°F				59	15	

Nonferrous Metals and Plastics 6

The ferrous metals, particularly steel and gray iron, hold such a predominant place in the economy that, for discussion, metals are usually divided into ferrous and nonferrous groups. On either a weight or a volume basis, pig iron is the cheapest refined metal form available today. Consequently, the use of nonferrous metals can generally be justified only on the basis of some special property that ferrous metals do not have or some processing advantage that a nonferrous metal offers.

Many Nonferrous Metals Exhibit Property Values. Nonferrous metals have a number of property advantages over steel and cast iron, although not all nonferrous metals have all the advantages. Aluminum, magnesium, and beryllium (one of the more rare metals) have densities of from one-fourth to one-third that of steel. Although strength rather than weight is more frequently the basis of design, in many cases, particularly in casting, the process limits the minimum section thickness, and products made of ferrous metals are made much stronger than required by the design. The same product made from even a weaker but less dense nonferrous metal may still have adequate strength and weigh much less. Even though the per pound cost of the nonferrous metal may be greater, the final costs of the products may be comparable. On a strength-to-weight basis, hardened steel is still superior to all but a few very high cost nonferrous metals, but some nonferrous alloys of only slightly less strength per unit weight may offer much greater ductility than the hardened steel and may be processed more economically. For the alloys shown in Figure 6-1, those classed as light alloys have one-fourth to one-third the density of iron or steel. Those called heavy alloys have densities approximately one to one and one-half times that of steel.

Corrosion Resistance Usually High. The corrosion resistance of most nonferrous metals is generally superior to all ferrous metals except stainless steel, and stainless steel does not offer the cost advantage of plain carbon and low alloy steels. This increased corrosion resistance is the most frequent reason for the choice of nonferrous metals.

Corrosion resistance is important for a number of reasons. Not only may the mechanical properties of the material be affected by corrosion but also the appearance of a metal is dependent on its corrosion resistance. Where appearance is important, the commonly used ferrous metals nearly always require some kind of finishing and protective surface treatment. With many nonferrous metals, protective finishes are not needed, even under conditions that would be severely corrosive to steel. The distinctive appearance of many nonferrous metals is highly desirable in many products.

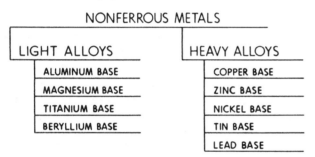

Figure 6-1
Nonferrous metals

Nonferrous Metals Used for Alloying with Iron as Well as Themselves. Although iron is the most frequently used magnet material, having high permeability and low magnetic hysteresis, pure iron is a poor permanent magnet material. The best permanent magnets are alloys high in nickel, aluminum, and cobalt. Silver, copper, and aluminum have much greater electrical and thermal conductivities than any ferrous materials and are usually used instead of steel when these properties are important.

Zinc Used in Large Quantities. Zinc is a typical example of a metal whose use in relatively large tonnages depends not so much on mechanical properties, or even on superior corrosion resistance, but on a special processing advantage. Zinc is weak, costs over twice as much per pound as pig iron or low carbon steel, and even with good corrosion resistance usually needs plating for good appearance, but its low melting point permits its use in die casting with longer die life than any other commonly cast metal.

ALUMINUM ALLOYS

Aluminum and copper are the most important of the nonferrous metals, being produced in approximately equal tonnages. However, about three-fourths of all the copper produced is used for electrical conductors, so aluminum is left as the most important structural nonferrous metal. Aluminum is potentially very available. Large ore deposits are found at many places, but the most economical reduction process yet developed still requires 8 kilowatt hours of electrical energy per pound of metal refined. Even so, the only cheaper metals on a weight basis are lead, zinc, and iron. Lead is seldom used as a structural metal, and zinc is limited mostly to low strength applications, so aluminum is a principal competitor with iron and steel. On a volume basis, only iron is cheaper.

GENERAL PROPERTIES

Strength of Aluminum Alloys. Aluminum alloys have tensile strengths that range from 83 to 550 MPa (12,000 to 80,000 psi). These values compare favorably with other nonferrous alloys and with many steels, although some steels may have strengths as great as 2,070 MPa (300,000 psi). Nevertheless, the low density of aluminum, about one-third that of iron, steel, and brass, is more important than space considerations.

Aluminum Has Excellent Ductility and Corrosion Resistance. The excellent ductility of aluminum permits it to be readily formed into complicated shapes and allows plastic flow instead of fracture failure under shock and other overload conditions. Pure aluminum has excellent corrosion resistance but is limited in use to those applications in which strength requirements are low. The corrosion resistance of the high strength aluminum alloys is generally good except when exposed to some alkaline environments. Additional protection may be provided for these conditions by cladding the alloys with a thin layer of the pure metal or other aluminum alloys.

When called upon to perform conductivity checks on aluminum plates and sheets, NDT personnel should be alert for clad materials. Since the electrical conductivities of the base metal and the cladding are invariably different, the eddy current conductivity measurement may include some combination of the two conductivities and result in misleading readings.

Some Poor Properties Restrict Use. The endurance limit even for hardened alloys is in the range of 5,000 to 20,000 psi. This weakness prohibits the use of aluminum in some applications in which vibration is combined with high stress levels, and it is often necessary to observe special precautions to eliminate the occurrence of stress risers, such as notches, scratches, and sudden section changes. Another deficiency is the loss of strength that occurs with increased temperature. Both work-hardened and heat-treat-hardened alloys lose strength rapidly at temperatures greater than about 150° C. This loss of strength at elevated temperatures not only restricts the design of parts made of aluminum but also, because it is combined with a loss of ductility near the melting point (a condition called *hot shortness*), makes the processes of casting and welding more difficult.

Aluminum Alloys Provide Valuable Combined Properties. All of the metals and alloys, both ferrous and nonferrous, have some combination of properties that make them preferred for some applications. While aluminum is exceeded in any individual property by some other metal and while it has deficiencies that limit its use, the combination of properties it possesses (particularly good corrosion resistance, conductivity, lightness, good strength-to-weight ratio, and good ductility), when combined with easy fabrication and moderately low cost, account for its importance as a structural metal second only to iron and steel.

WROUGHT ALUMINUM ALLOYS

Uses for All Pure Metals Limited. Aluminum alloys designed to be used with some deformation process, in which ductility and strain-hardening properties are of greatest importance, are referred to as wrought alloys. Any pure metal, including aluminum, generally has greater ductility, higher conductivity, and better corrosion resistance than any alloyed form of the metal. The purest readily available form of aluminum has especially high conductivity and is designated as electrical grade (EC). Compared to copper, its conductivity is 68% on a volume basis but 200% on a weight basis.

Pure Aluminum — Soft and Weak but Corrosion Resistant. Highest purity is necessary only for electrical use. Commercially pure aluminum has sufficient impurities present to impair its electrical conductivity significantly but retains excellent corrosion resistance and ductility. In the fully softened condition the tensile strength is about 83 MPa (12,000 psi). When fully work hardened, the strength is approximately doubled. The combination of high ductility and low strength generally results in poor machinability, particularly from the standpoint of surface finish.

Pure Aluminum and Most Alloys Not Hardenable by Heat Treatment. Neither electrical grade nor commercially pure aluminum is susceptible to hardening by heat treatment. Likewise, a number of aluminum alloys containing alloying elements that remain in solid solution at all temperatures do not respond to heat-treat-hardening procedures. The effect of the alloys is to increase the strength at the expense of some ductility. Tensile strengths in the range of 110 to 275 MPa (16,000 to 40,000 psi) when annealed and 50% to 70% greater when fully work hardened may be obtained by additions of manganese, chromium, magnesium, and iron. Alloys of this type offer advantages over pure aluminum by compromising with a reduction of forming properties to gain in mechanical properties. The additional strength is obtained only by the presence of the alloy in solid solution and not because of heat treatment. However, the alloys are subject to *work hardening* and *recrystallization* treatments as are all metals. The term *annealing*, when used with reference to pure aluminum or one of the solid solution alloys, can only be interpreted to mean recrystallization.

PROPERTY CHANGES

Hardening and Strengthening by Heat Treatment. As was discussed in Chapter 4, the possibility of heat-treat hardening exists even in metals that undergo no allotropic changes when an alloying element is more soluble at elevated temperature than at room temperature. Varying amounts of copper, less than about 5.5%, can be alloyed with aluminum. Depending on the heat treatment, three different structures may be actually obtained. With slow cooling, equilibrium conditions are approximated, and the alloy is placed in its softest, or annealed, condition. With rapid cooling, the fully saturated structure of intermediate hardness and ductility is obtained. Following the establishment of this supersaturated structure, the alloy is subject to aging, either natural with time at room temperature or artificial at slightly elevated temperature. The hardest structure is obtained only by heating to the solution temperature to allow the copper to form a solid solution, followed by quenching and aging.

Reaction with Magnesium or Magnesium-Silicon Similar to Copper. Besides copper-aluminum, two other reactions of this type are used in commercial aluminum alloys. Above 4%, magnesium forms heat treatable alloys with aluminum, and the combination of magnesium and silicon forms the compound MgSi, which acts in the same way as copper or pure magnesium. Strengths of these alloys range from 90 to 241 MPa (13,000 to 35,000 psi) in the annealed condition and from 241 to 550 MPa (35,000 to 80,000 psi) in the fully hardened condition.

The fabricator of aluminum products may obtain the alloys in a number of different heat-treated and work-hardened conditions. Table 6-1 shows the standard symbols that are used to denote these conditions. The terms *solution treated, aged, annealed,* and *cold worked* have been discussed in connection with heat treatments.

Slight Overaging Used for Stabilization. *Stabilizing* is an additional treatment used with aluminum alloys to control growth and distortion. In an alloy naturally or artificially aged to the maximum hardness level, a period of time follows during which the natural relieving of stresses will result in uncontrolled, though small, dimensional changes. If the aging process is carried slightly past that required for maximum hardness, the structure is dimensionally *stabilized* and no further significant changes will occur.

TABLE 6-1
Aluminum temper and heat-treat symbols

−F		As fabricated.
−O		Annealed (recrystallized) temper of wrought materials.
−H	1	Strain hardened only. Degree of hardening designated by second digit 1 through 8. Second digit 9 used to designate extra hard temper.
−H	2	Strain hardened and partially annealed. Second digit 2 through 8 used in same manner as for H1 series.
−H	3	Strain hardened and stabilized. Second digit to designate degree of residual strain.
−T	2	Annealed temper of cast material.
−T	3	Solution treat and strain harden.
−T	4	Solution treat and natural age.
−T	5	Artificial age only after cooling from elevated processing temperature.
−T	6	Solution treat and artificial age.
−T	7	Solution treat and stabilize.
−T	8	Solution treat, strain harden, and artificial age.
−T	9	Solution treat, artificial age, and strain harden.

Note: The above symbols follow the number designating the aluminum alloy type and become part of the material identification.

NDT Used for Conductivity Testing. Eddy current conductivity testing methods are in routine use for heat treatment control and alloy sorting. As shown in Table 6-2, the electrical conductivities of some common aluminum alloys vary, in some cases substantially, depending upon the variaions in alloying elements and heat treatment. However, examination of the % IACS values shown in Table 6-2 shows that in some cases the values are the same or nearly the same for different alloys. When sorting alloys by eddy current methods, it may be necessary to conduct a second test to positively identify the alloy or heat treatment condition. Frequently used for this purpose are chemical spot tests. While more difficult to use and not strictly nondestructive, chemical spot tests will conclusively identify the difference between, say 5052-0 and 2017-T4, which have electrical conductivities of 35 and 34% IACS respectively, a difference not conclusively separable by eddy current tests.

CAST ALUMINUM ALLOYS

Special Alloys Needed for Casting. Aluminum castings could be made from any of the alloys intended for plastic deformation. These alloys do in fact have their beginning as cast ingots, but certain alloys have been developed specifically for castings. As a cast metal, pure aluminum is subject to the same drawbacks that are characteristic of the wrought alloys. In addition, the relatively high melting temperature leads to excessive oxidation and the entrapment of gases in the molten metal. The fluidity of some liquid alloys is too poor for flow into thin sections, and some are subject to high shrinkage and cracking while solidifying and cooling in the mold. By proper alloying, all these conditions may be improved.

As with the wrought alloys, some casting alloys may be heat-treat hardened and some may not. Principal among the casting alloys that are not heat treatable are those containing silicon only. Used in amounts up to 11%, silicon improves fluidity and decreases shrinkage. Tensile strengths of 130 MPa (18,000 psi) for sand castings and 210 MPa (30,000 psi) for die castings are typical. Added magnesium improves not only the casting characteristics but also the machinability of the cast metal.

Heat-Treat-Hardenable Alloys More Difficult to Cast. Alloys subject to hardening by heat treatment are produced when copper only, magnesium plus silicon, or copper plus magnesium plus silicon are used as alloying elements. When subjected to a complete solution, quenching, and aging heat treatment, alloys of these types may have strength as great as 330 MPa (48,000 psi). Many aluminum castings are made of heat-treatable alloys and are used as cast, without heat treatment. The casting process itself generally provides rapid enough cooling to constitute a degree of quenching sufficient to give some supersaturation, and natural aging will provide some hardening. The use of the heat-treatable casting alloys is restricted to applications requiring high strength-to-weight ratios because these alloys are somewhat more difficult to cast. Shrinkage is generally higher than with the nonhardenable types, and the metal is more subject to cracking and tearing during the cooling period in the mold. (Table 6-2 shows some typical wrought and cast alloys.)

COPPER ALLOYS

While the total tonnage of copper has not decreased, the importance of this metal relative to ferrous metals and to other nonferrous metals has decreased throughout recent history. However, copper is the metal that has been of greatest importance during the longest period of man's history. The Bronze Age refers to the period of history during which man fashioned tools from copper and copper alloys as they were found to occur naturally in the free state. The copper used today is reduced from ores as are other metals, and the continued use depends on the properties that make it useful as either a pure or an alloyed metal.

TABLE 6-2
Some Aluminum Alloys and Properties

Type	Composition					Ten St. 1000 psi (6.9 × 10⁶ Pa)	Percent Elong (2 in.)	Hardness Brinell	Electrical Conductivity at 28° (% IACS)	Characteristics and Uses
	Cu	Si	Mn	Mg	Other					
						Wrought				
EC					.055	12 (O) 27 (H19)	23 (O) 1.5 (H19)		62 62	Good electrical conductor. Work harden only. Electrical conductors.
1100					1.0	13 (O) 24 (H18)	35 (O) 5 (H18)	23 (O) 44 (H18)	59 57	Good formability. High corrosion resistance. Work harden only. Cooking utensils, chemical equipment, reflectors.
3003			1.2			16 (O) 29 (H18)	30 (O) 4 (H18)	28 (O) 55 (H18)	50 40	Good corrosion resistance. Slightly less ductility. Work harden only. Extrusions, forgings, hardware.
5052				2.5	Cr 0.25	28 (O) 42 (H38)	25 (O) 7 (H38)	47 (O) 77 (H38)	35 35	Good corrosion resistance. Good machinability. Work harden only. Good weldability. Truck bodies, kitchen cabinets.
2017	4.0	0.7		0.5		26 (O) 62 (T4)	22 (O) 22 (T4)	45 (O) 105 (T4)	50 34	Corrosion resistance to rural atmosphere; poor corrosion resistance to marine atmosphere. Machinability good when hard. Screw machine products.
6061	0.28	0.6		1.0	Cr 0.25	18 (O) 45 (T6)	25 (O) 12 (T6)	30 (O) 95 (T6)	47 43	Excellent corrosion resistance to rural atmosphere; good resistance to industrial and marine atmosphere. Good weldability. Structures, marine use, pipes.
7075	1.6			2.5	Zn 5.6 Cr 0.29	33 (O) 83 (T6)	17 (O) 11 (T6)	60 (O) 150 (T6)	57 33	Good corrosion resistance to rural atmosphere but poor for others. Frequently clad. Good machinability. Poor weldability. Aircraft structure.
43		5.25				19 as cast	8	40		Good corrosion resistance. Only fair machinability. Sand and permanent mold castings, marine fittings, thin sections.
214				4.0		25 as cast	9	50		Good corrosion resistance. Excellent machinability. Sand castings, dairy and food-handling hardware.
355	1.25	5.0		0.5		42 Sol Treat Age	4	90		Good corrosion resistance. Good machinability. Sand and permanent mold castings.

GENERAL PROPERTIES

Copper is one of the heavier structural metals with a density about 10% greater than that of steel. Tensile strengths range from 210 to 880 MPa (30,000 to 125,000 psi), depending on alloy content, degree of work hardening, and heat treatment. The ductility is excellent and most alloys are easy to work by deformation processes, either hot or cold. The machinability ranges from only fair for some of the cast materials to excellent for some of the wrought materials. The most machinable are those containing lead or tin additives for the purpose of improving machinability.

Copper Has Excellent Thermal and Electrical Properties. If the preceding properties were the only properties of note that copper had, it would probably be little used. However, copper has outstanding electrical and thermal conductivity and excellent corrosion resistance, particularly when compared to ferrous metals. As noted before, three-fourths of the copper produced is used in pure form because of its conductivity. While aluminum has higher conductivity than copper on a weight basis and is displacing copper for some electrical applications, copper continues to be the principal metal for electrical use. This is particularly due to the higher strength-to-weight ratio of copper in pure-drawn form as is generally used for electrical conductors.

Corrosion Resistance to Some Environments Good. For other than electrical use, copper and its alloys compete with steel primarily because of better corrosion resistance. Copper alloys have excellent resistance to atmospheric corrosion, particularly under marine conditions. The combination of corrosion resistance and high thermal conductivity makes them useful for radiators and other heat exchangers.

BRASSES AND BRONZES

Definitions. For many years copper alloys were rather simply divided into two groups. Those containing zinc as the principal alloying element were known as *brass* and those containing tin as the principal alloying element were known as *bronze*. More recently the names have become confusing. Brasses generally contain from 5% to 40% zinc, but even one of these alloys is known as "commercial bronze." Bronzes contain a principal alloying element other than zinc. Tin is still the most common.

Properties Inverse to Cost. While the conductivity and ductility of any alloy is less than that of pure copper, strength, corrosion resistance to some media, machinability, appearance or color, and casting properties may be improved by alloying. Pure zinc is cheaper than pure copper, and the cost of their alloys becomes lower as the amount of zinc is increased. Bronzes generally have better properties than brasses, but the high cost of tin has limited their use. The low friction and excellent antiwear properties of bronze makes it preferred for many journal-bearing applications.

Some Alloys Heat-Treat Hardenable. A few of the copper alloys are hardenable by a solution-precipitation treatment similar to that used for aluminum. However, the high cost has confined their use to applications in which the combinations of high strength with high corrosion resistance or high strength with high conductivity are necessary. Most interesting of these alloys is one containing 98% copper and 2% beryllium. After proper heat treatment, this alloy has a tensil strength of 1,280 MPa (185,000 psi) and a Rockwell C hardness of 40. It is useful not only for applications such as electrical relay springs in which high endurance limit must be combined with high conductivity but also for chisels, hammers, and other tools for use in mines and other hazardous locations where sparks must be avoided.

Tin and Lead Improve Machinability and Castability. Copper alloys intended for casting usually contain some tin and lead to improve machinability and to reduce void formation in the castings. The properties that make the wrought alloys useful apply also to the cast alloys so that a large number of small castings are used in plumbing fixtures, marine hardware, pump impellers and bodies, electrical connectors, and statuary. Table 6-3 gives the compositions and properties of some typical brasses and bronzes.

NICKEL ALLOYS

Considerable Nickel Used as an Alloy in Steel. Nickel and manganese are metals that have mechanical characteristics similar to those of iron. However, neither is subject to alloying with carbon and control of hardness by heat treatment as is steel. Also, the ores of both metals are much less plentiful than iron ore, and the price is therefore higher. While manganese is little used except as an alloying element, nickel has sufficiently better corrosion and heat resistance than iron or steel to justify its use when these qualities are of enough importance. Nearly three-quarters of all the nickel produced is used either as a plating material for corrosion resistance or as an alloying element in steel. However, its use in steel has decreased in recent years with the discovery that other elements in lower percentages may have the same effects as nickel.

Most Important Property Is Corrosion Resistance. As a structural metal by itself, or as the basis of alloys, the properties of nickel and its alloys are indicated in Table 6-4. Nickel and copper are completely soluble in the solid state, and many different compositions are available. Those rich in copper compete with brass but have higher cost, corrosion resistance, and temperature resistance. Those richer in nickel have superior heat and corrosion resistance at even higher cost and are used in many applications in which stainless steel is used. The composition of Monel metal is determined largely by the composition of the ores found in the Sudbury district of Canada.

TABLE 6-3
Properties of some brasses and bronzes

Name	Composition Zn	Composition Sn	Composition Other	Ten. St 1000 psi (6.9 × 10⁶ Pa)	Percent Elong	Characterisitics and Uses
Electrolytic copper		99.9 pure		32-50	6-45	Excellent workability. Good electrical properties and corrosion resistance. Electrical conductors, contacts, switches, automobile radiators, chemical equipment.
Commercial bronze	10.0			37	5-45	Good corrosion resistance. Excellent workability. Marine hardware, costume jewelry.
Red brass	15.0			39-70	5-48	Good corrosion resistance to atmosphere. Good workability. Fair machinability. Weatherstrips, heat exchangers, plumbing.
Yellow brass	35.0			46-74	8-65	Good corrosion resistance to atmosphere. Poor hot workability. Fair machinability. Grillwork, lamp fixtures, springs.
Naval brass	39.0	1.0		25-53	20-47	Corrosion resistance generally good. Seldom used full hard. Aircraft and marine hardware, valve stems, condenser plates.
Phosphor bronze		10.0		66-128	3-68	Good atmospheric corrosion resistance. Poor machinability. Good wear qualities. Bearing plates, springs.
Aluminum bronze			7.0 Al	85-90	20-40	Excellent atmospheric corrosion resistance. Excellent hot workability. Good machinability. Gears, nuts, bolts.
Beryllium copper			0.3 Co 2.0 Be	185	3-50	Good atmospheric corrosion resistance. Poor machinability. Good workability. Springs, valves, diaphragms, bellows.

TABLE 6-4
Properties of some nickel alloys

Name	Mn	Fe	Cu	Other	Ten. St 1000 psi (6.9 × 10⁶ Pa)	Percent Elong	Characteristics and Uses
A Nickel	0.25	0.15	0.05		55-130	55-2	Corrosion-resistant at high temperature. Vacuum tube parts, springs, chemical equipment.
Monel	0.90	1.35	31.5		70-140	50-2	Good corrosion resistance combined with high strength at normal and medium temperatures. Pump shafts, valves, springs, food-handling equipment.
Inconel	0.20	7.20	0.10	Cr 15	80-170	55-2	Similar to Monel but better high temperature strength.
Nickel 36		64.0			70-90	36-20	Corrosion-resistant to atmospheres and to salt water. Low thermal expansion. Length standards, thermostatic bimetals.
Cast Monel	0.75	1.5	32.0	Si 1.6	65-90	50-20	Good corrosion resistance to salt water and most acids. Valve seats, turbine blades, exhaust manifolds.

(Composition Balance Nickel)

MAGNESIUM ALLOYS

Although beryllium is the lightest metal available, its extremely high cost restricts its use to very special applications. Magnesium is therefore the lightest metal commercially available, with a density two-thirds that of aluminum. Magnesium alloys have good strength, ranging up to 350 MPa (50,000 psi) for wrought alloys and up to 280 MPa (40,000 psi) for cast alloys. Corrosion resistance is good in ordinary atmosphere but for more severe conditions, including marine atmospheres, some surface protection is necessary.

Wrought and cast alloys have similar compositions. Aluminum, zinc, and manganese improve strength and forming properties. With 8% or more aluminum, a solution-precipitation hardening treatment is possible. Thorium, zirconium, and certain rare earth elements produce alloys useful at temperatures up to 480° C (900° F).

Magnesium Alloys Work Harden Easily. The principal drawbacks of magnesium, other than the relatively high cost of recovery from sea water, are related to its crystalline structure. Magnesium is one of the few important metals having a close-packed hexagonal structure. Characteristic of these metals is a high rate of strain hardening. This property has two practical consequences. The amount of cold working that can be done without recrystallization is quite limited so that most forming operations must be done hot. This causes no great difficulty in rolling, forging, and extrusion operations that are normally performed hot with any metal, but secondary press operations on flat sheet may require heating of the dies and magnesium sheet. Most pressworking equipment is not designed for this type of operation.

Stress Levels High at Notches and Imperfections. The high rate of strain hardening also results in the fault called *notch sensitivity*. At a stress concentration point, such as the base of the notch in an impact test specimen, the load-carrying ability of a material depends on its ability to permit some plastic flow to enlarge the radius and relieve the stress concentration. The high rate of strain hardening in magnesium lessens its ability to do this and thus lowers its impact test values, and makes it subject to failure at such imperfections as grinding marks, small shrinkage cracks from welding or casting, or sharp internal corners permitted as part of a design. For this reason, magnesium components used in aircraft and similar applications are inspected nondestructively usually by radiography for internal defects and by penetrant testing for surface defects.

Fine or Thin Magnesium Can Burn Readily in Air. Some problems are introduced in the processing of magnesium because of its inflammability. Reasonable care is necessary to prevent the accumulation of dust or fine chips where they might be subject to ignition from sparks, flames, or high temperatures.

ZINC ALLOYS

Low Cost but also Low Strength. Zinc has the lowest cost per pound of any nonferrous metal. However, its use has been restricted by a combination of factors. The best alloy has a tensile strength of 325 MPa (47,000 psi) and an endurance limit of 55 MPa (8,000 psi). When combined with the density, which is about the same as that of iron, the result is a fairly low strength-to-weight ratio. The recrystallization temperature of about 20° C for pure zinc simplifies tooling and drawing operations but results in a very low creep strength. Precipitation reactions with even small amounts of iron, lead, cadmium, or tin present as impurities can result in gradual dimensional change and loss of shock resistance with time. With artificial aging, the precipitation reaction that takes place with copper can be used to improve the mechanical properties of some alloys. Because of these limitations, zinc alloys are seldom used in a critical application that would warrant NDT.

Preferential Corrosion Feature Valuable. In addition to the relatively low cost, zinc has a number of other properties that make it desirable. It has good corrosion resistance when used as a coating on ferrous materials. The zinc is attacked in preference to the base metal, even though there are interruptions in the coating. Plating or coating with zinc is called *galvanizing* and accounts for the use of more than 50% of all the zinc produced.

Pure or slightly alloyed wrought zinc has high formability. It is an excellent roofing material and is frequently used for flashing on roofs of other materials. Its chemical properties make it useful for dry cell battery cases and for photoengraving plates.

Low Melting Temperature of Benefit for Die Casting. As a structural material, zinc is used almost exclusively because of its excellent casting properties in metal molds. With pouring temperatures ranging from 740° to 800° C, zinc alloys used in die casting give much greater die life than magnesium, aluminum, or copper alloys. The higher density of zinc than of aluminum or magnesium is offset to some extent in die casting by the fact that zinc can be die cast in thinner sections than other metals. Although zinc has good natural corrosion resistance, this property can be improved along with appearance by appropriate platings, which are easy to apply.

SPECIAL GROUPS OF NONFERROUS ALLOYS

HEAT- AND CORROSION-RESISTANT ALLOYS

Several different groups of materials, some including certain ferrous alloys, have traditionally been grouped on the basis of property requirements rather than base metal or alloy content. Of special importance and increasing interest recently have been alloys designed for use under high stress conditions at elevated temperatures in such applications as jet turbine engines, high temperature steam piping and boilers, and rocket combustion chambers and nozzles. The efficiency of many such devices depends on the maximum temperature at which they can be operated, and they frequently involve highly oxidizing, corrosive, or erosive conditions.

Manufacturing Cost High. Most special materials that have been developed for these uses are difficult to process into usable products by some or all of the standard procedures. The high cost of such products is due both to the generally high cost of the materials themselves (rarity and cost of refining) and the cost of special processing. Hot working involves extra high temperatures with high forces, which results in short equipment life; casting frequently must be done by investment or other high cost techniques; cold working is difficult or impossible; welding involves elaborate procedures to avoid contamination and nondestructive testing to insure reliability; and machining requires low cutting speeds with short tool life even under the best conditions.

Stainless Steels. These alloys may be divided into three rather roughly defined groups. Stainless steels, which were discussed earlier, have better strength and corrosion resistance than plain carbon or low alloy steels at temperatures higher than 1,200° F. A number of alloys of the same general composition as standard stainless steels have been developed with larger amounts of nickel and generally larger amounts of the stabilizing elements such as titanium or molybdenum for better high temperature properties. Aluminum or copper may be used to provide a precipitation reaction that makes the alloys hardenable by heat treatment. Such heat treatment usually involves solution temperatures higher than 1,000° C and artificial aging at temperatures higher than 700° C.

Nickel Alloys. Nickel-based alloys form a second group of high temperature materials. They normally contain chromium or cobalt as the principal alloying element and smaller amounts of aluminum, titanium, molybdenum, and iron. These alloys have better properties at high temperatures than the stainless steel types but cost more and are even more difficult to process.

Cobalt Alloys. Alloys having cobalt as the principal element form a third group. They are generally referred to as cobalt-based alloys, although they may not contain as much as 50% of any single element. Other elements are generally nickel, chromium, tungsten, columbium, manganese, molybdenum, and carbon. Alloys of this type are useful structurally at temperatures as high as 1,000° C, at which they have good corrosion resistance and tensile strengths as great as 90 MPa (13,000 psi).

OTHER NONFERROUS METALS

Of the many other potential base metals, most are used under special conditions. Many of these metals have properties that are equal to or better than those of iron and the more common nonferrous metals, but their use is restricted by economic consideration. Gold, platinum, and other *noble* metals have high chemical inertness, but their rarity and high cost restrict their use. Beryllium has the highest strength-to-weight ratio of any known metal, but the difficulty of obtaining the pure metal and the rarity of the ore make the cost almost as high as that of gold. Titanium ores are abundant and titanium has extremely useful properties, but the cost of reduction is approximately one hundred times that of iron. Titanium could easily be the most important nonferrous metal if low cost production methods could be developed. Table 6-5 gives the principal characteristics and uses for most nonferrous metals that are available commercially.

NON-METALS

PLASTICS

For some time, the fastest growing field of materials has been the group called plastics. Any thorough treatment of plastics, especially concerning the chemistry of the materials, would require a number of volumes. On the other hand, plastics cannot properly be ignored in any treatment of materials and manufacturing processes because they are in direct competition with most metals. Since 1958, a greater tonnage of plastics has been produced annually than of all nonferrous metals combined.

Many Materials — Wide Range of Properties. A study of plastics is complicated by the tremendous number of material variations possible. There are roughly as many important families of plastics as there are commercially important metals. While it is true that many of the metals are alloyed to different combinations, the number is relatively small when compared with the number of distinct plastics possible in each family. Furthermore, while for metals the hardness and strength seldom exceeds a ratio of perhaps 10:1 for any particular alloy group, many plastics that are under a single name are produced with properties ranging from liquids that are used as adhesives or finishes to rigid solids whose hardness and strength compare favorably with metals.

TABLE 6-5
Characteristics of most nonferrous metals

Metal	Principal Characteristics	Applications	
		Pure or as Base Metal	As Alloying Constituent
Antimony	Hard, brittle	None	1%-12% hardens lead fusible alloys, 2% hardens copper
Beryllium	Lightest structural metal, High strength/weight ratio, Brittle, transparent to X-rays	Aircraft and rocket structure, X-ray tube windows	
Bismuth	Soft, brittle, high negative coefficient of resistivity	Use restricted by cost Special resistance elements	Fusible alloys
Cadmium	Higher temperature strength than tin- or lead-based alloy, corrosion-resistant	Plating, especially on steel; bearing alloys; solders	Bearing alloys, solders
Cerium	Soft, malleable, ductile	Rare	Lighter flints, nodular iron Raises creep strength of Al Mg, Ni, Cr
Cobalt	Weak, brittle, high corrosion-resistant	Rare	High temperature alloys, permanent magnets, hard facing tool steels
Columbium (Niobium)	High melting point, corrosion-resistant	Nuclear reactors, missiles, rockets, electron tubes	High temperature alloys, stainless steels, nitriding steels
Germanium	Brittle, corrosion-resistant, semiconductor	Diodes, transistors	Rare
Gold	Ductile, malleable, weak, corrosion-resistant	Monetary standard, plating, jewelry, dental work, electrical contacts	Rare
Indium	Soft, low melting point	None	Hardener for silver and lead. Corrosion resistance in bearings.
Iridium	Most corrosion-resistant metal	None	Hardener for platinum jewelry, contact alloys
Lead	Weak, soft, malleable, corrosion-resistant	Chemical equipment, storage batteries, roof flashing, plumbing	Improves machinability of steel and most nonferrous alloys, solders, bearing alloys
Manganese	Moderate strength, ductile	Rare	To 2% low alloy steels, 12% abrasion-resistant steel, stainless steels
Mercury	Liquid at room temperature	Thermometers, switches	Low melting point alloys Amalgam with silver for dental use
Molybdenum	High melting point, high strength at elevated temperature, oxidizes rapidly at high temperature	High temperature wire, structural use with surface protection, mercury switch contacts	Low alloy steels, high temperature alloys, stainless steel, tool steels
Palladium	Ductile, corrosion-resistant	Chemical catalyst, electrical contacts	Jewelry, dental alloys
Rhodium	High reflectivity, free from oxidation films, chemically inert	Mirrors, plating	With platinum and palladium
Selenium	Special electrical and optical properties	Rectifiers, photocells	Machinability of stainless steel
Silver	Highest electrical conductivity, corrosion resistance to nonsulphur atmospheres	Coinage, jewelry, tableware, electrical contacts, plating, catalyst, reflectors	Brazing and soldering alloys, bearing alloys
Silicon	Semiconductor, special electrical and optical properties	Rectifiers, transistors, photocells	Electrical steel, cast iron, cast nonferrous

TABLE 6-5—Continued

Metal	Principal Characteristics	Applications Pure or as Base Metal	As Alloying Constituent
Tantalum	High melting point, ductile, corrosion-resistant	Surgical implants, capacitors, chemical hardware, electronic tubes	Tantalum carbide cutting tools
Tin	Soft, weak, malleable, corrosion resistant	Plating, collapsible tubes	Bronzes, solders, bearing alloys
Titanium	Density between steel and light alloys, high strength, corrosion-resistant	Marine, chemical, food-processing equipment Aircraft, rockets, orthopedic and orthodontic equipment	High temperature alloys, stainless steel, aluminum alloys, titanium carbide tools
Tungsten	Highest melting point of metals; strong, high modulus of elasticity; corrosion-resistant	Lamp filaments, contacts, X-ray targets, nuclear reactors	Alloy steels, tool steels, high temperature alloys, tungsten carbide tools
Vanadium	Moderate strength, ductile	Rare	Alloy steel, tool steel, nonferrous deoxidizer
Zirconium	Moderate strength, ductile, corrosion-resistant	Structural parts in nuclear reactors	Stainless steels

Definition Difficult. The word plastic is derived from the Greek word *plastikos*, which meant "fit for molding." Many of the materials called plastics today, such as finishes and adhesives, are not molded at all; moreover, many materials are molded that are not called plastics. Many metals and most ceramics are molded at times. Plastics might best be defined as a group of large-molecule organic compounds, primarily produced as a chemical product and susceptible to shaping under combinations of pressure and heat. To include all the plastics, the term organic must be expanded to include silicone-based as well as carbon-based, materials.

Major Development Recent. Historically, the development of plastics has occurred in two general periods. Chemists in France, Germany, and England, during the period from 1830 to 1900, isolated and named many materials that are called plastics today. The actual commercial production of most of these materials was delayed until production methods and facilities became available that permitted them to compete with the more traditional materials. The second period of even more rapid developments has been in the United States, particularly since 1940. Many new methods of manufacture and treatment as well as new plastic materials have been developed.

PLASTIC MATERIALS

Plastic Structure. Chemically, plastics are all *polymers*. The smallest unit structure, or molecule, that identifies the chemical involved is called a *monomer*. By various means, including heat, light, pressure, and agitation, these monomers may be made to join and grow into much larger molecules by the process of *polymerization*. In general, the first polymerization involves the connecting of the monomers into long chains, usually with a progressive degree of solidification or an increase in viscosity as the polymerization proceeds. For most plastics, the properties depend on the degree of polymerization, which explains to a large degree the wide range of properties available. For the group of plastics known as *thermosetting*, a second type of polymerization takes place in which cross-linking occurs between adjacent chains. This thermosetting reaction frequently results in greatly increased rigidity.

TYPES OF PLASTICS

Long Chain Polymers. There are two broad groups of plastics, based originally on their reaction to heat but more properly on the type of polymerization involved. Plastics that are called *thermoplastic* have the degree of polymerization controlled in the initial manufacture of the plastic raw material, or *resin*. These materials soften with increasing temperature and regain rigidity as the temperature is decreased. The process is essentially reversible, but in some cases, chemical changes that may cause some deterioration of properties are produced by heating.

Thermosetting Plastics — Cross-Linked Polymers. As noted before, the thermosetting plastics undergo a further cross-linking type of polymerization, which for the early plastics was initiated by the application of heat, but which for many modern thermosetting plastics may be initiated by other means. In the fabrication by molding of thermosetting plastics, an initial thermoplastic stage is followed by the thermosetting reaction at higher temperatures or with prolonged heating. Thermoplastics may be resoftened by reheating, but the thermosetting reaction is chemical in nature and irreversible so that once it has taken place, further heating results only in gradual charring and deterioration.

The origin of the resin distinguishes a number of different types of plastics. Some true plastics are found in nature and used essentially as found. These include shellac, used most frequently as a finish for wood and as an adhesive constituent, and asphalt, used as a binder in road materials, as a constituent in some finishes, and, with fibrous filling materials, as a molding compound.

Some Plastics — Natural Materials. A number of plastics are natural materials that have undergone some chemical modification but retain the general chemical characteristics of the natural material. Cellulose may be produced as paper with slight modification, as vulcanized fiber with a slightly greater modification, and as cellulose acetate with even more modification. Wood in its natural state has thermoplastic properties that are used in some manufacturing processes. Rubber latex, as found in nature, is a thermoplastic material but is generally modified by chemical additions to act as a thermosetting material.

Most Plastics — Synthetic. The greatest number of plastics presently used are most properly called *synthetic* plastics. While many of them make use of some particular natural material, such as petroleum, as the principal constituent, the chemistry of the raw material and the chemistry of the finished plastic have no direct connection. The raw material may be thought of simply as the source of elements and compounds for the manufacture of the plastic.

CHARACTERISTICS OF PLASTICS

Tables 6-6 and 6-7 give the principal characteristics and typical uses for most of the plastics in common use. No such list can be complete because new plastics are constantly being introduced, and the time span from discovery of a useful plastic to commercial use is decreasing. The cellulose plastics among the thermoplastics and phenol formaldehyde (a phenolic) among thermosetting plastics were the first plastics to be developed and are still in wide use today.

General Property Comparisons. Some comments may be made about the chart, keeping in mind that most general rules have exceptions. As a group, thermoplastics are somewhat lower in strength and hardness but higher in toughness than thermosetting materials. The thermosetting plastics generally have better moisture and chemical resistance than the thermoplastics. The terms *high* and *low*, when used for strengths, service temperatures, and other characteristics, are only relative and apply to plastics as a total group.

None of the plastics have useful service temperatures that are as high as those of most metals, and the modulus of elasticity of all plastics is low compared to most metals. While the ultimate strengths of many metals are greater than that available with plastics, some specific plastics offer favorable comparisons. Nylon, for example, is one of a few plastics that, being truly crystalline, may be hardened by working. Drawn nylon filaments may have a tensile strength of 50,000 psi, which is actually greater than some low strength steels. Plastics excel in some applications as insulators or where chemical resistance is important. The greatest tonnage, however, is used in direct competition with other materials where plastics may be favored because of their low fabrication costs in large quantities, light weight, and easy colorability.

TABLE 6-6
A summary of principal characteristics and uses of thermoplastic plastics

Resin Type	Principal Characteristics	Forms Produced	Typical Uses	Relative Cost
ABS	High strength, toughness, colorability	Injection moldings, extrusions, formable sheet	Pipe, appliance cabinets, football helmets, handles	50-60
Acetal	High strength, colorability, high fatigue life, low friction, solvent resistance	Injection moldings, extrusions	Gears, impellers, plumbing hardware	80
Acrylic	High strength, colorability, optical clarity, low service temperature	Injection moldings, extrusions, castings, formable sheet, fiber	Transparent canopies, windows, lenses, edge-lighted signs, mirrors, high quality molded parts	45-55
Cellulose acetate	Moderate strength, toughness colorability, optical clarity, wide hardness range, low service temperature	Injection moldings, extrusions, formable sheet, film, fiber	Toys, shoe heels buttons, packaging, tape	36-58
Cellulose acetate butyrate	Moderate strength, high toughness, good weatherability, colorability, optical clarity, low service temperature	Injection moldings, extrusions, formable sheet, film	Telephone handsets, steering wheels, appliance housings, outdoor signs, pipe	40-62
Cellulose propionate	Moderate strength, high toughness, good weatherability, colorability, optical clarity, low service temperature	Injection moldings, extrusions, formable sheet, film	Radio cabinets, pen and pencil barrels, automobile parts	40-62
Ethyl cellulose	Moderate strength, high toughness, flexibility, colorability, moisture resistance, better electric properties than other cellulostics, low service temperature	Injection moldings, extrusions, film	Refrigerator parts, aircraft parts, flash-light housings, door rollers	65-75
Cellulose nitrate	Toughest of all thermoplastics, good formability, poor aging, high flammability, low service temperature	Extrusions, formable sheet	Ping-Pong balls, hollow articles	70-200

TABLE 6-6—Continued

Chlorinated polyether	High chemical resistance, moderate strength	Injection moldings, extrusions, sheet	Valves, pump parts in corrosive environments	250
TFE (tetrafluoroethylene)	Chemical inertness, high service temperature, low friction, low creep strength, high weatherability	Sintered shapes, extrusions, formable sheet, film, fiber	Pipe, pump parts, electronic parts, nonlubricated bearings, gaskets, antiadhesive coatings	350-550
CFE (chlorotrifluoroethylene)	Higher strength than TFE, lower chemical resistance than TFE, high service temperature, high weatherability	Injection moldings, extrusions, formable sheet, film	Coil forms, pipe, tank lining, valve diaphragms	700-800
Nylon (polyamide)	High strength, toughness, work hardenability, low friction, good dielectric properties	Injection moldings, extrusions, formable sheet, film, fiber	Gears, cams, bearings, pump parts, coil forms, slide fasteners, gaskets, high pressure tubing	100-200
Polycarbonate	High strength, toughness, chemical resistance, weatherability, high service temperature	Injection moldings, extrusions	Gears, hydraulic fittings, coil forms, appliance parts, electronic components	150
Polyethylene	Moderate strength, high toughness, good dielectric properties, low friction, chemical resistance, flexibility	Injection moldings, extrusions, formable sheet, film, fiber, rigid foam	Housewares, pipe, pipe fittings, squeeze bottles, sports goods, electrical insulation	32-38
Polystyrene	High strength, low impact resistance, high dielectric strength, colorability, optical clarity, low service temperature	Injection moldings, extrusions, formable sheet, film, foam	Toys, electrical parts, battery cases, light fixtures, rigid conduits	22-43
Vinyl	Wide range of properties, strength, toughness, abrasion resistance, colorability, low service temperature	Compression moldings, extrusions, castings, formable sheet, film, fiber, foam	Electrical insulation, floor tile, water hose, raincoats	24-43

TABLE 6-7
A summary of principal characteristics and
uses of thermosetting plastics

Resin Type	Principal Characteristics	Forms Produced	Typical Uses	Relative Cost
Epoxy	Moderate strength, high dielectric strength, chemical resistance, weatherability, colorability, high service temperature, strong adhesive qualities	Casting, reinforced moldings, laminates, rigid foam, filament wound structures	Chemical tanks, pipe, printed circuit bases, short-run dies, randomes, pressure vessels	45-80
Melamine	Hardest plastic, high dielectric strength, moderate service temperature, colorability, dimensional stability	Compression and transfer moldings, reinforced moldings, laminates	Dinnerware, electrical components, table and counter tops	42-45
Phenolics	Moderately high strength, high service temperature, dimensional stability, color restrictions	Compression and transfer moldings, castings, reinforced moldings, laminates, cold moldings rigid foam	Electrical hardware, poker chips, toys, buttons, appliance cabinets, thermal insulation, table and counter tops, ablative structural shapes	20-45
Polyester (including alkyds)	Moderately high strength, dimensional stability, fast cure, easy handling, good electrical properties, high service temperatures, chemical resistance	Castings, reinforced moldings, laminates, film, fiber, compression and transfer moldings	Electrical parts, automobile ignition parts, heater ducts, trays, tote boxes, laundry tubs, boats, automobile bodies, buttons	31-60
Silicon	Highest service temperatures, low friction, high dielectric strength, flexible, moderate strength, high moisture resistance	Compression and transfer moldings, reinforced moldings, laminates, rigid foam	High temperature electrical insulation, high temperature laminates, gaskets, bushings, seals, spacers	275-540
Urea	Moderately high strength, colorability, high dielectric strength, water resistance, low service temperature	Compression and transfer moldings	Colored electrical parts, buttons, dinnerware	19-34
Urethane	Moderate strength, high toughness, very flexible, colorable, good weatherability excellent wear resistance, low service temperatures	Injection moldings extrusions, blow moldings, foam	Gears, bearings, O-rings, footwear, upholstery foam	50-100

The Nature of Manufacturing 7

The height reached and the progress made by any past civilization is judged by many factors. In some cases a civilization is most remembered for cultural advances in the areas of art and literature. More commonly, however, the degree of advancement is measured by the quality and quantity of durable goods produced. The use of the terms *Stone Age*, *Bronze Age*, and *Iron Age* is based on the extent of man's knowledge and ability in the areas of materials and processing during these periods of history. A similar situation continues today. The United States is envied throughout the world for its ability to produce and distribute durable goods in large quantities.

Regardless of whether or not it is justified, present-day evaluations of individuals, organizations, and countries are most frequently based on the goods used by them. Even the production of food is dependent on the manufacture of modern farm machinery and chemicals.

Specialization — a Basis for Progress. Early man must have been faced with many problems. Even as in some areas of the world today, he must have spent the major portion of his time in satisfying basic needs for food and shelter. All he had for tools and raw materials were those that were immediately at hand. It is reasonable to assume, however, that even at a bare subsistence level, some men were better food gatherers than others, some were better weapon makers, and some were better cave diggers. While the transition from an individual existence to one of specialization undoubtedly occurred only after long periods of time, it is the idea of specialization that has been basic to man's progress throughout history.

System Control Essential. Obviously, if each worker performs where he is best qualified, the overall work efficiency will be high and the product output maximum for any given technology. Such specialization, however, can be accomplished only under some organizational control. For individuals to be willing to become specialists, it is essential that a system of exchange and distribution be established. A control system of some type is necessary to balance the various specialty outputs and to set the values of service and product output.

The control organization may be based on undelegated authority (master and slave), on delegated authority (elected officials), or on natural controls as the result of supply and demand in a free enterprise system. Our system today is based on the latter two in which elected officials and supply and demand are the principal controls.

MODERN MANUFACTURING

MARKETS

Manufacturing in any period of history has been characterized by certain essential features. One requirement is that a market exist for any goods produced. A natural market exists for those things that are deemed essential to life, such as food, but for most manufactured goods, a market must be created by a requirement that is sometimes based on an expected standard of living rather than on any basic biological need of man.

Product Life Usually Limited. Few durable goods have truly unlimited life. Because of the economics of manufacturing or the requirements of a design, the life of most products is limited. For all practical purposes, an automobile body made of titanium would have unlimited life so far as corrosion is concerned, but the cost would be prohibitive for a mass market, and the life of the body would be limited by design changes and wearout of other parts of the automobile. The blades in a jet turbine have limited life, not because it is desirable, but because of design considerations of weight and the limited properties of the available materials.

In addition to wearout, a market for replacement exists because of obsolescence. New designs, new materials, or new features may make replacement desirable either for convenience, as with many new automobiles, or for economic reasons. Machine tools are generally replaced while they are still in working condition, but their replacement is justified on the basis of lower maintenance, higher productivity, and higher accuracies of newer designs.

Product Markets Grow for Several Reasons. For nearly all durable goods, growth has created an expanding market in the world and especially in the United States. This growth has occurred in two forms. Not only has the population been continually increasing but also the rising standard of living has made a greater percentage of the population able to buy most durable consumer goods. More leisure time and increased purchasing power have caused large increases in the sale of many products. This has been particularly noticeable in automobiles, housing, household appliances, and recreational equipment. A part of the increase in per capita consumption must be attributed to the improved sales and advertising techniques that have developed in this country based to a large extent on the better communications that exist today.

The greatest increase in markets, particularly in the last 50 years, is due to new inventions and to new applications of older products and materials. In many cases the new products have been made economically possible primarily by improved processing machinery and techniques. Many of the currently used plastics have been known to chemists for over a hundred years, but the development of the plastics industry to its current state depended on the development of economical methods of raw material production and fabrication as well as on the development of a product demand.

Market Forecasting Difficult but Essential. Other new products are based on basic concepts or discoveries that did not exist 50 years ago. The whole electronics industry, especially that depending on the transistor and solid-state physics, falls in this category. The increased complexity, cost, and specialization of modern industry have led to an increased need for knowledge of expected demand prior to the time sales are actually made. In the production of goods sold seasonally, it is necessary that the proper inventories be built up with a relatively constant level of production or that the work be balanced with other products in order that plant investment may be kept reasonable. Accurate forecasts of future demand are essential when increases in plant capacity or new plants for the production of new goods are anticipated, for the investment in a single new plant may be over $100 million.

DESIGN

Appearance in Addition to Function Usually Important. In the case of every product, the manufacturing process must be preceded by the design. The relationships that exist between design and processing are of extreme importance. The designer normally starts with some definite functional requirement that must be satisfied. The environmental conditions of use, expected life, and loading conditions will dictate certain minimum shapes and sizes and limit the possible choice of materials. The designer's problems arise mainly from the fact that a single solution is seldom indicated. Of the many possible materials and shapes that may satisfy the functional requirements, some may have better appearance than others. For many consumer goods, the appearance may actually govern the final choice. Even in the designing of parts that may be completely hidden in a final assembly, the designer seldom disregards appearance completely.

Quality and Costs Must Balance. Even the original design will be influenced by the method of processing that is anticipated and, to give proper consideration to all the alternatives, it is essential that the designer have knowledge about the costs and capabilities of various production methods. It is generally true that costs will be different for different material and processing choices, and considerable screening of the alternatives can be done purely on a cost basis. However, the quality obtained with more expensive materials or methods may be superior to that of the cheaper choices, and decisions must often be made regarding some combination of quality and cost. A rational decision as to the quality to be produced can only be made with adequate information as to how the market will be affected by the quality.

Availability of Facilities Affects Choices. Obviously, the decisions made by the designer are far reaching and of extreme importance. The materials and shapes that he specifies usually determine the basic processes that must be used. Tolerances that he specifies may even dictate specific types of machines and will have a large influence on costs. In many cases his choices are limited by the equipment and the trained personnel that are available. Economical manufacture of small quantities can frequently be best accomplished by use of equipment and processes that under other circumstances would be inefficient. Certainly a designer for a plant producing castings would not design a part as a weldment if the continued operation of the plant depends on the production of castings.

In many cases, the decisions that govern the choice of materials and processes must be made in an arbitrary manner. The gathering of enough information may not be economically feasible, or time may not be available. Particularly when only small quantities are being produced, the cost of finding the most economical method of production may be more than any possible gain over some arbitrary method that is reasonably certain of producing an acceptable product. In some cases, custom governs the choice simply because some set of choices was known to give acceptable results for similar production in the past.

Designers cannot be expected to be experts in all the phases of production that influence the final quality and cost of a product. Production personnel must be relied on to furnish details of process capabilities and requirements.

NDT in Design. Similarly, the design engineering function must receive technical guidance from key NDT personnel in order to assure that the design requirements can be met. It is essential that the design requirements contain the proper balance between the contribution from NDT to safety and reliability of the product and the economic realities. Both the capabilities and limitations of the various methods of NDT must be considered in the design phases of the product life cycle in order to achieve optimum product effectiveness.

PROCESSING

Manufacturing Usually a Complex System. While the problems of design and processing are interrelated, once the design decisions have been made, the problems of processing are more clearly defined. A design may indicate certain processing steps, but basically the problem in processing is to make a *product* whose material, properties, shapes, tolerances, size, and finish meet specifications laid down by the designer. Manufacturing is a term usually used to describe that section of processing starting with the raw material in a refined bulk form, and is concerned mainly with *shape changing*. While the single operation of sawing to length might produce a product useful as fireplace wood, for most manufactured products of metals, plastics, and other materials, a complex series of shape- or property-changing steps is required.

The Usual Processing Steps for Metals. Figure 7-1 shows the basic processes that are used in shaping metals. The reduction of ores is essential to any further processing, and the choices in processing come later. All but a very small percentage of the metal that is refined is first cast as a pig or ingot, which is itself always the raw material for further processing.

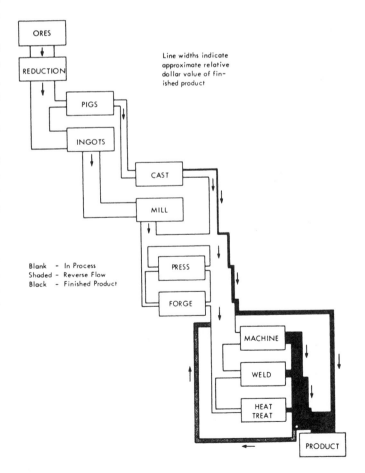

Figure 7-1
Metals process flow

It can be noted, however, that from this point on, any process may either produce a finished product or furnish the raw material for some further processing. The reverse flow shown in the lower part of the diagram refers particularly to parts that have been heat treated or welded and must then be machined. This step generally would occur only once for any product.

It is the rule rather than the exception, however, that many reversals may occur within some of the blocks on the diagram. Steel is commonly subjected to several different rolling operations in a steel mill. Pressworking operations most often involve several separate steps to produce a product. The greatest amount of repetition occurs in machining. It is not unusual for a complex part, such as an automobile engine block, to be subjected to as many as eighty separate machining operations.

The majority of manufacturing organizations specialize in one type of manufacturing operation, and even the extremely large companies that may operate in several fields of manufacturing generally have specialized plants for the separate manufacturing areas.

STATES OF MATTER

Material may exist in one of three states of matter, gas, liquid, or solid, but except for some special processes with relatively small use, such as vapor deposition, or for zinc refining, the gaseous state is of small importance in manufacturing.

Most Manufacturing Processes Are to Change Material Shapes. For manufacturing purposes in which shape changing is the objective, the solid state may be thought of as existing in two forms. Below the elastic limit, materials are dealt with as rigid materials. Processing involving this form causes no significant relative movement of atoms or molecules of the material with respect to each other. Above the elastic limit, solid materials may flow plastically, and shape changing may be accomplished by application of external loads to cause permanent relocations within the structure of the material. The end results of dealing with materials in the liquid form are similar to those with materials above the elastic limit. No appreciable density or volume change occurs, and the shape may be changed without loss of material.

SHAPE-CHANGING PROCESSES

Shapes Changed by No Volume Change, by Additions, and by Subtractions. Shape changing is possible in any of these states, but most manufacturing processes by definition or nature deal with materials in only one of these possible forms. Figure 7-2 shows the processes for shape changing without material loss and those in which material is added or taken away.

No Volume Change. In those processes in which no volume change occurs, property changes are usually large and distributed throughout the material. In casting, the shape change occurs by melting and subsequent solidification to a prescribed shape. This process can be used with practically all metals and most plastics. The material properties depend on composition and the conditions of the particular casting process, but not on the condition of the material prior to melting. Casting is often the most economical method for producing complex shapes, particularly where reentrant angles exist.

Wrought materials are produced by plastic deformation that can be accomplished by hot working (above the recrystalization temperature) or cold working. Property changes also occur throughout the material with these processes; the greatest changes are usually caused by cold work.

Additions or Combinations. New shapes can be produced either by joining preformed shapes mechanically or by any of various bonding means. In welding, soldering, and brazing, metallurgical bonds are established by heat, pressure, or sometimes by chemical action with plastics. Mechanical fastening by use of bolts, rivets, or pins is primarily an assembly procedure and is often an alternative and competitive joining procedure to welding or adhesive fastening.

Shaping from powders by pressing and heating involves the flow of granular materials, which differs considerably from deformation processing, although some plastic flow undoubtedly occurs in individual particles. Powder processing is a somewhat specialized process, but, as in casting or the deformation processes, the material is shaped by confinement to some geometric pattern in two or three dimensions. Because the total volume of work material is affected by these processes, large sources of energy, pressure, or heat are required.

Subtraction or Removal. Shape changing may also be accomplished by taking material away in chip or bulk form or by material destruction. The property changes in these processes are more localized, and energy requirements are generally smaller.

Mechanical separation can be performed by removal of chips or by controlled separation along predetermined surfaces. Chip removal by machining can be used with some success for all materials, shapes, and accuracies and is probably the most versatile of all manufacturing processes. Separation by shearing, with localized failure caused by externally applied loads, is limited primarily to sheet materials but frequently turns out to be the cheapest method for producing many shapes in large quantities.

Special Shape-Changing Methods. Particularly in recent years, with the advent of new materials difficult to fabricate by conventional means and of many

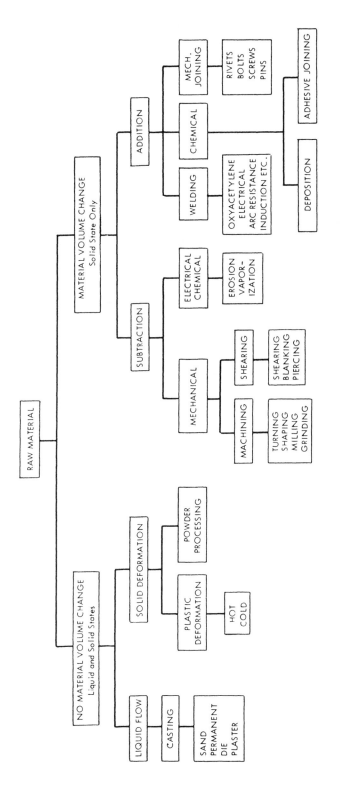

Figure 7-2
Shape-changing processes

designs requiring shapes and tolerances and material combinations difficult to achieve with conventional processes, a number of electrical and chemical processes have been developed for removing or adding material. Many of these are restricted in use to a few materials, and most are specialized to the point that they have only a few applications. Included is metal plating by electrical or chemical means, used primarily as a finishing process. Other developments are electrical discharge "machining," chemical milling, ultrasonic grinding, and electron beam machining, which are specialized metal removal processes that compete with conventional machining or press-working operations and involve hard materials, special shapes, or low quantities.

SUMMARY

Manufacturing is a complex system. A product always originates as a design concept required to serve some purpose. A multiplicity of choices and decisions nearly always comes between the establishment of the need and the manufacturing of the product. The designer, because no logical means are available, frequently arbitrarily makes decisions that usually, at least broadly, determine the processes which must be used to produce the product. Within this broad framework, however, exist many other choices of specific materials, processes, and machines. Materials, properties, qualities, quantities, and processes are strongly interrelated. The prime effort, from original concept to the completion of manufacture, is aimed at finding the optimum combination of these variables to provide the best economic situation.

Since NDT is an inseparable part of the manufacturing system, it is imperative that NDT personnel in responsible positions must have general knowledge of the elements of manufacturing technology. The NDT specialist will devote many hours in analysis and interpretation of the flaws and faults resulting from manufacturing operations. In order to provide input to corrective action, he will be called upon many times to furnish technical guidance to the design, materials, manufacturing and quality assurance functions. Without some knowledge of the total manufacturing process, the NDT specialist cannot adequately fulfill these responsibilities.

The Casting Process   8

Casting is the process of causing liquid metal to fill a cavity and solidify into a useful shape. It is a basic method of producing shapes. With the exception of a very small volume of a few metals produced by electrolytic or pure chemical methods, all material used in metal manufacturing is cast at some stage in its processing. Castings of all kinds of metals, in sizes from a fraction of an ounce up to many tons, are used directly with or without further shape processing for many items of manufacture. Even those materials considered to be wrought start out as cast ingots before deformation work in the solid state puts them in their final condition.

A vast majority of castings, from a tonnage standpoint, are made from cast iron. A relatively small number of these are subjected to NDT. In most cases they are designed for non-critical applications with principally compressive loading and oversize dimensions to eliminate the problem effect of the innumerable discontinuities inherent in the material. However, some of these castings and many others made of different material may be used in such a way that careful inspection is essential for satisfactory service. Penetrant testing may be in order for surface examination. Radiography or ultrasonic testing may be needed to detect internal defects regardless of the material or type of casting. Ultrasonic methods are difficult to use with some castings because of noise created by grain structure. The rough surfaces of many castings also can produce problems in transducer coupling, but ultrasonic testing is used extensively in the examination of critical coolant passages in turbine engine blades to measure thickness. Eddy current and penetrant methods are also used to detect leading and trailing edge cracks before and during service of turbine blades.

THE PROCESS

The Process Starts with a Pattern. The casting, or founding, process consists of a series of sequential steps performed in a definite order, as shown in Figure 8-1. First, a pattern to represent the finished product must be chosen or constructed. Patterns can be of a number of different sytles, but are always the shape of the finished part and roughly the same size as the finished part with slightly oversized dimensions to allow for shrinkage and additional allowances on surfaces that are to be machined. In some casting processes, mainly those performed with metal molds, the actual pattern may be only a design consideration with the mold fulfilling the function of a negative of the pattern as all molds do. Examples would be molds for ingots, die casting, and permanent mold castings. Most plastic parts are made in molds of this type, but with plastics, the process is often called molding rather than casting.

A Mold Is Constructed from the Pattern. In some casting processes, the second step is to build a mold of material that can be made to flow into close contact with the pattern and that has sufficient strength to maintain that position. The mold is designed in such a way that it can be opened for removal of the pattern. The pattern may have attachments that make grooves in the mold to serve as channels for flow of material into the cavity. If not, these channels, or *runners*, must be cut in the mold material. In either case, an opening to the outside of the mold, called a *sprue*, must be cut or formed.

Mold Cavity Filled with Molten Material. Liquid metal is poured through the channels to fill the cavity completely. After time has been allowed for solidification to occur, the mold is opened. The product is then ready for removing the excess metal that has solidified in the runners, cleaning for removal of any remaining mold material, and inspecting to determine if defects have been permitted by the process. The casting thus produced is a finished product of the foundry. This product occasionally may be used in this form, but more often than not needs further processing, such as machining, to improve surface qualities and dimensions and, therefore, becomes raw material for another processing area.

Casting Is a Large Industry. The tonnage output of foundries throughout the United States is very large, consisting of close to 20 million tons (18 million tonnes) per year. Foundries are scattered all over the United States but are concentrated primarily in the eastern part of the nation with a secondary concentration on the west coast in the two areas where the main manufacturing work is carried on.

Foundries Tend to Specialize. Because of differences in the problems and equipment connected with casting different materials, most foundries specialize in producing either ferrous or nonferrous castings. Relatively few cast both kinds of materials in appreciable quantities in the same foundry.

A few foundries are large in size, employing several thousand men, but the majority are small with from one to one hundred employees. Most large foundries are captive foundries, owned by parent manufacturing companies that use all, or nearly all, of the foundry's output. More of the small foundries are independently owned and contract with a number of different manufacturers for the sale of their castings. Some foundries, more often the larger ones, may produce a product in sufficient demand that their entire facility will be devoted to the making of that product with a continuous production-type operation. Most, however, operate as job shops that produce a number of different things at one time and are continually changing from one product to another, although the duplication for some parts may run into the thousands.

SOLIDIFICATION OF METALS

The casting process involves a change of state of material from liquid to solid with control of shape being established during the change of state. The problems associated with the process, then, are primarily those connected with changes of physical state and changes of properties as they may be influenced by temperature variation. The solution to many casting problems can only be attained with an understanding of the solidification process and the effects of temperature on materials.

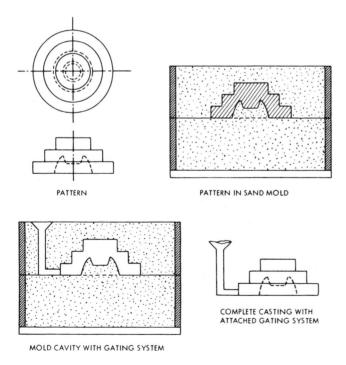

Figure 8-1
Casting steps for a pulley blank

SOLIDIFICATION

Energy in the form of heat added to a metal changes the force system that ties the atoms together. Eventually, as heat is added, the ties that bind the atoms are broken, and the atoms are free to move about as a liquid. Solidification is a reverse procedure, as shown in Figure 8-2, and heat given up by the molten material must be dissipated. If consideration is being given only to a pure metal, the freezing point occurs at a single temperature for the entire liquid. As the temperature goes down, the atoms become less and less mobile and finally assume their position with other atoms in the space lattice of the unit cell, which grows into a crystal.

Crystal Growth Starts at the Surface. In the case of a casting, the heat is being given up to the mold material in contact with the outside of the molten mass. The first portion of the material to cool to the freezing temperature will be the outside of the liquid, and a large number of these unit cells may form simultaneously around the interface surface. Each unit cell becomes a point of nucleation for the growth of a metal crystal, and, as the other atoms cool, they will assume their proper position in the space lattice and add to the unit cell. As the crystals form, the heat of fusion is released and thereby increases the amount of heat that must be dissipated before further freezing can occur. Temperature gradients are reduced and the freezing process retarded. The size of crystal growth will be limited by interference with other crystals because of the large number of unit cell nuclei produced at one time with random orientation. The first grains to form in the skin of a solidifying casting are likely to be of a fine equiaxed type with random orientation and shapes.

Second Phase Slower. After formation of the solid skin, grain growth is likely to be more orderly, providing the section thickness and mass are large enough to cause a significant difference in freezing time between the outside shell and the interior metal. Points of nucleation will continue to form around the outside of the liquid as the temperature is decreased. The rate of decrease, however, continues to get lower for a number of reasons. The heat of fusion is added. The heat must flow through the already formed solid metal. The mold mass has been heated and has less temperature differential with the metal. The mold may have become dried out to the point that it acts as an insulating blanket around the casting.

Second Phase Also Directional. Crystal growth will have the least interference from other growing crystals in a direction toward the hot zone. The crystals, therefore, grow in a columnar shape toward the center of the heavy sections of the casting. With the temperature gradient being small, growth may occur on the sides of these columns, producing structures known as dendrites (Figure 8-3). This pine-tree-shaped first solidification seals off small pockets of liquid to freeze later. Evidence of this kind of crystal growth is often difficult to find when dealing with pure metals but, as will be discussed later, can readily be detected with most alloy metals.

Third Phase. As the wall thickness of frozen metal increases, the cooling rate of the remaining liquid decreases even further, and the temperature of the remaining material tends to equalize. Relatively uniform temperature distribution and slow cooling will permit random nucleation at fewer points than occurs with rapid cooling, and the grains grow to large sizes.

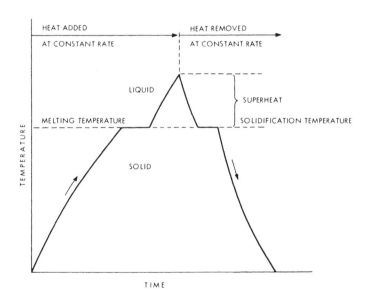

Figure 8-2
Heating and cooling curves for temperature increase above the melting point for a metal

Figure 8-3
Schematic sketch of dendritic growth

Grain Characteristics Influenced by Cooling Rates. As shown in Figures 8-4 and 8-5, it would be expected in castings of heavy sections that the first grains to form around the outside would be fine equiaxed. Columnar and dendritic structure would be present in directions toward the last portions to cool for distances depending upon the material and the cooling rate under which it is solidified. Finally, the center of the heavy sections would be the weakest structure made up of large equiaxed grains. Changes in this grain-growth pattern can be caused by a number of factors affecting the cooling rate. Thin sections that cool very quickly will develop neither the columnar nor the coarse structure. Variable section sizes and changes of size and shape may cause interference and variations of the grain-structure pattern. Different casting procedures and the use of different mold materials can affect grain size and shape through their influence on the cooling rate.

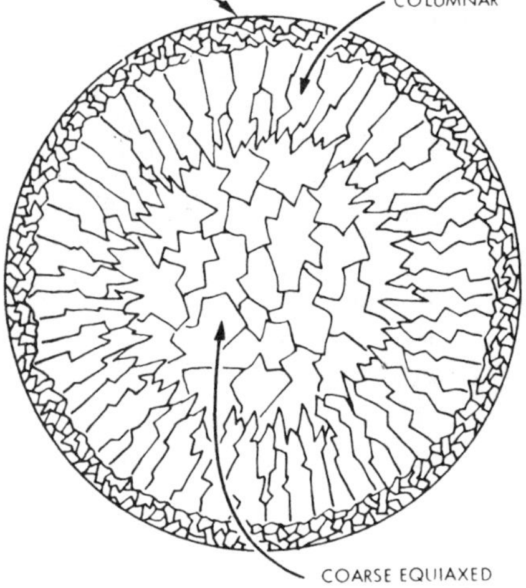

Results of NDT for internal defects may be difficult to analyze because of effects from variable grain size in massive castings. Large grains cause diffraction efects with radiographic methods and reflection from grain boundaries causes problems with ultrasonic testing. Special techniques which minimize these effects may be necessary to test large grained castings.

Figure 8-4
Typical grain structure from solidification of a heavy section

Eutectics Similar to Pure Metals. Eutectic alloys freeze in much the same manner as a pure metal. Solidification takes place at a single temperature that is lower than that for the individual components of the alloy. The grain size produced with an eutectic alloy is smaller than the grain size of a pure metal under the same conditions. It is believed that this is due to a smaller temperature gradient and the formation of a greater number of points of nucleation for the start of grains.

Noneutectics Freeze through a Temperature Range. The majority of products are made from noneutectic alloys. Instead of freezing at a single temperature as does the pure metal and the eutectic alloys, the noneutectic alloys freeze over a temperature range. As the temperature of the molten material is decreased, solidification starts at the surface and progresses toward the interior where the metal is cooling more slowly. Partial solidification may progress for some distance before the temperature at the surface is reduced low enough for full solidification to take place. The material at temperatures between those at which solidification begins and ends is partially frozen with pockets of liquid remaining to produce a mixture that is of mushy consistency and relatively low strength. Figure 8-6 is a graphic representation of this kind of freezing. The duration of this condition and the dimensions of the space between the start and finish of freezing are functions of the solidification temperature range of the alloy material and the thermal gradient.

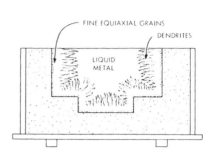

Figure 8-5
Grain formation in a heavy sand casting

The greater the solidification temperature range (in most cases meaning the greater the variation away from the eutectic composition) and the smaller the temperature gradient, the greater the size and duration of this mushy stage.

Segregation. Dendritic grain growth is much more evident in the noneutectic alloy metals than in pure metal. When more than one element is present, segregation of two types occurs during solidification. The first solids to freeze will be richer in one component than the average composition. The change caused by this *ingot-type segregation* is small, but as the first solids rob the remaining material, a gradual change of composition is caused as freezing progresses to the center. The other type of segregation is more localized and makes the dendritic structure easy to detect in alloy materials. The small liquid pockets, enclosed by the first dendritic solids, have supplied more than their share of one component to the already frozen material. This difference in composition shows up readily by difference in chemical reaction if the material is polished and etched for grain examination.

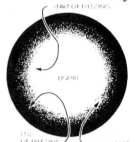

Figure 8-6
Process of freezing in a noneutectic alloy

SHRINKAGE

Shrinkage Occurs in Three Stages. Some of the most important problems connected with the casting process are those of shrinkage. The amount of shrinkage that occurs will, of course, vary with the material being cast, but it is also influenced by the casting procedure and techniques. The three stages of contraction that occur as the temperature decreases from the temperature of the molten metal to room temperature are illustrated in Figure 8-7.

First Stage Shrinkage in the Liquid. In the melting procedure, preparatory to pouring castings, the metal is always heated well above the melting temperature. The additional heat above that necessary for melting is called superheat. It is necessary to provide fluidity of the liquid to permit cold additives to be mixed with the metal before pouring. Superheat allows the metal to be transferred and to contact cold equipment without starting to freeze, and insures that sufficient time will elapse before freezing occurs to allow disposal of the material. Some superheat is lost during transfer of the liquid metal from the melting equipment to the mold. However, as the metal is poured into the mold, some superheat must remain to insure that the mold will fill. Loss of superheat results in contraction and increased density but is not likely to cause serious problems in casting. The volume change can be compensated for by pouring additional material into the mold cavity as the superheat is lost. An exception exists when the cavity is of such design that part of it may freeze off and prevent the flow of the liquid metal for shrinkage replacement.

Solidification Shrinkage. The second stage of shrinkage occurs during the transformation from liquid to solid. Water is an exception to the rule, but most materials are more dense as solids than as liquids. Metals contract as they change from liquid to solid. The approximate volumetric solidification shrinkage for some common metals is shown in Table 8-1. Contraction at this stage can be partially replaced

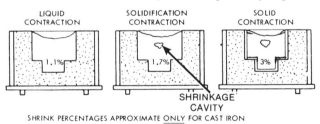

Figure 8-7
Three stages of metal contraction

because the entire metal is not yet frozen. If a suitable path can be kept open, liquid metal can flow from the hot zones to replace most of the shrinkage. It will be remembered, however, that in the formation of a dendritic grain structure, small pockets have been left completely enclosed with solid material. Depending upon the characteristics of the material and the size of the liquid enclosures, localized shrinking will develop minute random voids referred to as microporosity or microshrinkage (Figure 8-8). Micro-

TABLE 8-1
Approximate solidification shrinkage of some common metals

Metal	Percent Volumetric Shrinkage
Gray iron	0–2
Steel	2.5–4
Aluminum	6.6
Copper	4.9

porosity causes a reduction in density and tends to reduce the apparent shrinkage that can be seen on the surface of a casting.

The shrinkage that occurs during solidification and the microporosity that often accompanies it are minimized in materials that are near eutectic composition. This seems to be due to more uniform freezing with lower temperature gradients and more random nucleation producing finer grain structure. Microshrinkage is often a problem in aluminum or magnesium castings.

Macroporosity. The porosity of a casting may be amplified by the evolution of gas before and during solidification. Gas may form pockets or bubbles of its own or may enter the voids of microporosity to enlarge them. The evolved gas is usually hydrogen, which may combine with dissolved oxygen to form water vapor. These randomly dispersed openings of large size in the solid metal are referred to as macroporosity.

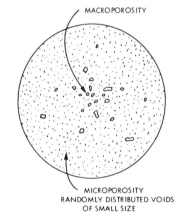

Figure 8-8
Porosity

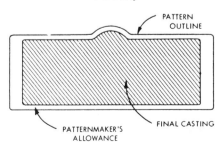

Figure 8-9
Pattern shrinkage allowance

Contraction in the Solid State. The third stage of shrinkage is that occurring after solidification takes place and is the primary cause of dimensional change to a size different form that of the pattern used to make the cavity in the mold. Although contraction of solidification may contribute in some cases, the solid metal contraction is the main element of *patternmaker's shrinkage*, which must be allowed for by making the pattern oversize.

POURING AND FEEDING CASTINGS

CASTING DESIGN

The first consideration that must be given to obtain good castings is to casting design. It should be remembered that although volumetric shrinkage of the liquid is thought of as being replaced by extra metal poured in the mold and by hydraulic pressure from elevated parts of the casting system, this can be true only if no parts of the casting freeze off before replacement takes place. Except for the small pockets completely enclosed by solid metal in the development of dendritic structures, the shrinkage of solidification can be compensated for if liquid metal can be progressively supplied to the freezing face as it advances.

Progressive versus Directional Solidification. The

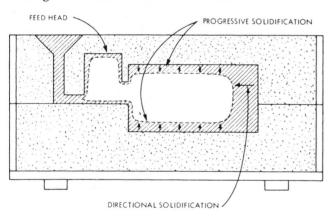

Figure 8-10
Progressive and directional solidification

term *progressive solidification*, the freezing of a liquid from the outside toward the center, is different from *directional solidification*. Rather than from the surface to the center of the mass, *directional solidification* is used to describe the freezing from one part of a casting to another, such as from one end to the other end, as shown in Figure 8-10. The direction of freezing is extremely important to the quality of a casting because of the need for liquid metal to compensate for the contraction of the liquid and that during solidification. Casting design and procedure should

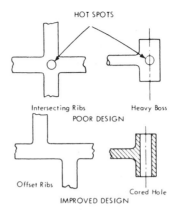

Figure 8-11
Hot spot elimination

cause the metal farthest from the point of entry to freeze first with solidification moving toward a *feed head*, which may be at the point where metal is poured into the mold or can be located at other points where liquid can be stored to feed into the casting proper.

Hot Spots Are Focal Points for Solidification. The highest temperature areas immediately after pouring are called *hot spots* and should be located as near as possible to sources of feed metal. If isolated by sections that freeze early, they may disturb good directional solidification with the result that shrinks, porosity, cracks, ruptures, or warping will harm the casting quality. It is not always necessary to completely inspect some castings when the vulnerable spots can be determined by visual inspection. Defects are most likely at hot spots created by section changes or geometry of the part and where gates and risers have been connected to the casting.

Control of Hot Spots Usually by Proper Design. Hot spots are usually located at points of greatest sectional dimensions. Bosses, raised letters, non-uniform section thicknesses, and intersecting members are often troublemakers in the production of high quality castings. Solution to the problem involves changing the design, as shown in Figure 8-11, or pouring the casting in such a way that these spots cease to be sources of trouble. Changing the design might include coring a boss to make it a thin-walled cylinder, relieving raised letters or pads on the backside, proportioning section thicknesses to uniform changes of dimensions, using thin-ribbed design instead of heavy sections, spreading and alternating intersecting members, and making other changes that will not affect the function of the part but will decrease the degree of section change.

Uniform Section Thicknesses Desirable. As a general rule, section changes should be minimized as much as possible in order to approach uniform cooling rates and reduce defects. When pouring iron, heavy sections tend to solidify as gray iron with precipitated graphite. Thin sections of the same material cooling at higher rates tend to hold the carbon in the combined state as iron carbide with the result that these sections turn out to be hard, brittle white iron. Since it is clearly impossible to design practical shapes without section changes, the usual procedure calls for gradual section-size changes and the use of liberal fillets and rounds. Some section changes are compared in Figure 8-12.

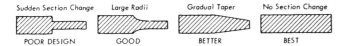

Figure 8-12
Section changes in casting design

POURING

Most Pouring Done from Ladles. Pouring is usually performed by using ladles to transport the hot metal from the melting equipment to the molds. Most molds are heavy and could be easily damaged by jolts and jars received in moving them from one place to another. Exceptions exist with small molds or with heavier molds, with which special equipment is used, that can be conveyorized and moved to a central pouring station. Even with these, the hot metal is usually poured from a ladle, though some high production setups make use of an automatic pouring station where spouts are positioned over the mold and release the correct amount of metal to fill the cavity.

Turbulent Flow Harmful. Casting quality can be significantly influenced by pouring procedure. Turbulent flow, which is caused by pouring from too great a height or by excessive rates of flow into the mold, should be avoided. Turbulence will cause gas to be picked up that may appear as cavities or pockets in the finished casting and may also oxidize the hot metal to form metallic oxide inclusions. Rough, fast flow of liquid metal may erode the mold and result in loss of shape or detail in the cavity and inclusion of sand particles in the metal. *Cold shots* are also a result of turbulent flow. Drops of splashing metal lose heat, freeze, and are then entrapped as globules that do not join completely with the metal which freezes later and are held partly by mechanical bond.

Pouring Rate. The pouring rate used in filling a mold is critical. If metal enters the cavity too slowly, it may freeze before the mold is filled. Thin sections that cool too rapidly in contact with the mold walls may freeze off before the metal travels its complete path, or metal flowing in one direction may solidify and then be met by metal flowing through another path to form a defect known as a *cold shut*. Even though the mold is completely filled, the cold shut shows the seam on the surface of the casting, and the metal is not solidly joined and is therefore subject to easy breakage.

If the pouring rate is too high, it will cause erosion of the mold walls with the resulting sand inclusions and loss of detail in the casting. High thermal shock to the mold may result in cracks and buckling. The rate of pouring is controlled by the mold design and the pouring basin, sprue, runner, and gate dimensions. The gating system should be designed so that when the pouring basin is kept full, the rest of the system will be completely filled with a uniform flow of metal.

Superheat Affects Casting Quality. As mentioned earlier, metals are superheated from 100° to 500° above their melting temperature to increase their fluidity and to allow for heat losses before they are in their final position in the mold. For good castings, the metal must be at the correct superheat at the time it is poured into the mold. If the temperature is too low, misruns and cold shuts will show up as defects in the casting, or the metal may even freeze in the ladle. If the temperature at pouring is too high, the metal may penetrate the sand and cause very rough finishes on the casting. Too high pouring temperatures may cause excessive porosity or increased gas development leading to voids and increased shrinkage from thermal gradients that disrupt proper directional solidification. High pouring temperature increases the mold temperature, decreases the temperature differential, and reduces the rate at which the casting cools. More time at high temperature allows greater gain growth so that the casting will cool with a weaker, coarse grain structure.

THE GATING SYSTEM

Metal is fed into the cavity that shapes the casting through a gating system consisting of a *pouring basin*, a *down sprue*, *runners*, and *ingates*. Some typical systems are shown in Figure 8-13. There are many special designs and terminology connected with these channels and openings whose purpose is that of improving casting quality. Special features of a gating system are often necessary to reduce turbulence and air entrapment, reduce velocity and erosion of sand, and remove foreign matter or dross. Unfortunately, no universal design is satisfactory for all castings or materials. There are no rules that can be universally

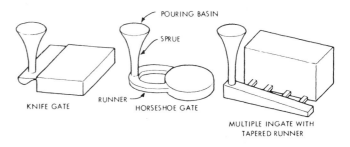

Figure 8-13
Typical gating systems

depended upon, and experimentation is commonly a requirement for good casting production.

The location of the connection for the gate, or gates, can usually be determined visually. These spots are possible concentration points for defects.

RISERS

Risers Are Multipurpose. Risers, feeders, or feed heads serve as wells of material attached outside the casting proper to supply liquid metal as needed to compensate for shrinkage before solidification is complete. Although most liquid contraction is taken care of during pouring, a riser may supply replacement for some of this contraction after parts of the casting have frozen solid, as shown in **Figure 8-14**. However, the principal purposes of risers are to replace the contraction of solidification and to promote good directional solidification. The need for risers varies with the casting shape and the metal being poured.

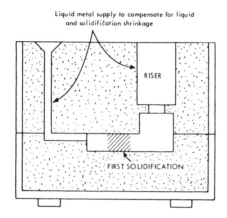

Figure 8-14
Risers for shrinkage control

CHILLS

Chills Initiate Solidification. Help in directional solidification can also be obtained in a reverse manner by the use of *chills*, which are heat-absorbing devices inserted in the mold near the cavity (Figure 8-15). To absorb heat rapidly, chills are usually made of steel, cast iron, or copper and designed to conform to the casting size and shape. Because chills must be dry to avoid blowhole formation from gases, it is sometimes necessary to pour a mold soon after it has been made, before the chills have time to collect moisture from condensation. In addition to helping with directional solidification, chills may also improve physical properties. Fast cooling during and after solidification retards grain growth and thus produces a harder, stronger structure.

Choice of Internal Chills Critical. Internal chills that become an integral part of the casting are occasionally used to speed solidification in areas where external chills cannot be applied. The design and use of internal chills is critical. Usually this type of chill is made of the same material as the casting. The chill must be of such size that it functions as a cooling device, but at the same time it must be heated enough that it fuses with the poured material to become an integral and equally strong part of the casting.

Nondestructive testing is often used to detect unfused internal chills and adjacent defects that may be caused by the change in cooling rate created by the presence of the chill.

FOUNDRY TECHNOLOGY

Although the casting process can be used to shape almost any metal, it has been necessary to develop a number of different methods to accommodate different materials and satisfy different requirements. Each method has certain advantages over the others, but all have limitations. Some are restricted to a few special applications.

SAND MOLDING

Sand is the most commonly used material for construction of molds. A variety of sand grain sizes, combined and mixed with a number of other materials and processed in different ways, causes sand to exhibit characteristics that make it suitable for several applications in mold making. A greater tonnage of castings is produced by sand molding than by all other methods combined.

Procedure for Sand Molding. The following requirements are basic to sand molding, and most of them also apply for the construction of other types of molds.

1. Sand — To serve as the main structural material for the mold
2. A pattern — To form a properly shaped and sized cavity in the sand
3. A flask — To contain the sand around the pattern and to provide a means of removing the pattern after the mold is made
4. A ramming method — To compact the sand around the pattern for accurate transfer of size and shape
5. A core — To form internal surfaces on the part (usually not required for castings without cavities or holes)

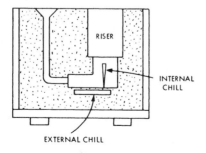

Figure 8-15
Chills as an aid to directional solidification

6. A mold grating system — To provide a means of filling the mold cavity with metal at the proper rate and to supply liquid metal to the mold cavity as the casting contracts during cooling and solidification

The usual procedure for making a simple green sand casting starts with placing the pattern to be copied on a *pattern*, or *follower*, board inside one-half of the flask, as shown in **Figure 8-16**. Sand is then packed around the pattern and between the walls of the flask. After striking off excess sand, a *bottom board* is held against the flask and sand and the assembly turned over. Removal of the pattern board exposes the other side of the pattern. A thin layer of *parting* compound (dry nonabsorbent particles) is dusted on the pattern and sand to prevent adhesion. Addition of the upper half of the flask allows sand to be packed against the pattern.

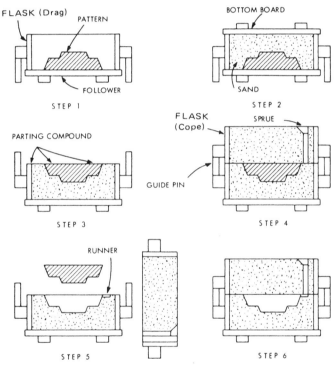

Figure 8-16
Principal steps for making a sand mold

After the sprue is cut to the parting line depth, the upper half of the mold can be removed, the pattern withdrawn, and the gating system completed. Reassembly of the mold halves completes the task, and the mold is ready for pouring.

GREEN SAND

The Word Green Refers to Moisture. The majority of castings are poured in molds of *green sand*, which is a mixture of sand, clay, and *moisture*. The materials are available in large quantities, are relatively inexpensive, and except for some losses that must be replaced, are reusable. The proportions of the mixture and the types of sand and clay may be varied to change the properties of the molds to suit the material being poured. To produce good work consistently, it is important that advantage be taken of the properties that can be controlled by varying the constituents of the sand mixture.

Sand Grains Held Together by Clay. In a mold, the sand particles are bound together by clay that is combined with a suitable quantity of water. The most commonly accepted theory of bonding is that as pressure is applied to the molding sand, clay, coating each sand particle, deforms and flows to wedge and lock the particles in place. The clay content can be varied from as little as 2% or 3% to as high as 50%, but the best results seem to be obtained when the amount of clay is just sufficient to coat completely each of the sand grains.

Water Conditions the Clay. Water is the third requisite for green sand molding. The optimum quantity will vary from about 2% to 8% by weight, depending largely upon the type and quantity of clay present. Thin films of water, several molecules in thickness, are absorbed around the clay crystals. This water is held in fixed relationship to the clay by atomic attraction and is described as rigid water, or tempering water. The clays that have the greatest ability to hold this water film provide the greatest bonding strength. Water in excess of that needed to temper the molding sand does not contribute to strength but will improve the flowability that permits the sand to be compacted around the pattern.

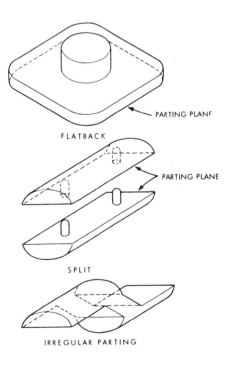

Figure 8-17
Common loose pattern types

PATTERNS

By most procedures, patterns are essential for producing castings. In occasional emergency situations an original part, even a broken or worn part, may be used as a pattern for making a replacement, but considerable care and skill is necessary when this is done.

Patterns are made of various materials: principally wood, metal, plastic, or plaster, depending on the shape, size, intricacy, and amount of expected use. They are constructed slightly larger than the expected resulting part to allow for shrinkage of the liquid metal, during and after solidification, to room temperature size. Extra matrial is also left on surfaces to be machined or finished to provide removal material on the casting. Patterns also must be contructed with suitable *draft* angles to facilitate their removal from the mold medium. Patterns may be designated as *flat-back* where the largest two dimensions are in a single plane, *split* which effectively separates to form flat-back patterns, or *irregular parting* which requires separation along two or more planes for removal of the pattern to produce the casting cavity. Any of these pattern types can be mounted on a matchplate for improved accuracy and faster production if justified by the needed quantity of castings. Some pattern types of the loose variety are shown in Figure 8-17.

FLASKS

Flasks are open faced containers that hold the molten medium as it is packed around the pattern. They are usually contructed in two parts: the upper half *cope* and the lower half *drag* (see Figure 8-16) which are aligned by guide pins to insure accurate positioning. The separation between the cope and drag establishes the parting line and when open permits removal of the pattern to leave the cavity whose walls form the casting when liquified material solidifies against it.

Some flasks, used most for small quantity casting, are *permanent* and remain around the sand until after pouring has been completed. Others used for higher production quantities are removable and can be used over and over for construction of a number of molds before pouring is required. The removable flasks are of three styles: *snap* flasks, having hinged corners, that can be unwrapped from the mold; *pop-off* flasks that can be expanded on two diagonal corners to increase the length and width to allow removal; and *slip* flasks that are made with movable sand strips that project inside to obstruct sliding of the mold medium until they are withdrawn to permit removal of the flask from the mold. When molds are constructed with removable flasks, *jackets* are placed over them to maintain alignment during pouring.

SAND COMPACTION

Casting Quality Dependent on Proper Compaction. Compaction, packing, or ramming of sand into place in a mold is one of the greater labor and time-consuming phases of making castings. It also has considerable influence on the quality of finished castings produced. Sand that is packed too lightly will be weak and may fall out of the mold, buckle, or crack, which will cause casting defects. Loosely packed grains at the surface of the cavity may wash with the metal flow or may permit metal penetration with a resulting rough finish on the casting. Sand that is too tightly compacted will lack permeability, restrict gas flow, and be a source of blowholes, or may even prevent the cavity from completely filling. Too tightly packed sand may also lack collapsability so that as solidification occurs, cracks and tears in the casting may be caused by the inability of the sand to get out of the way of the shrinking metal. Each of the several available methods for compacting sand has advantages over the others and limitations that restrict its use.

Butt Ramming Involves Human Effort. Peen and butt rammers may be used on a bench or on the floor by manual operation, or, in the case of large molds, the work may be done with pneumatic rammers similar to an air hammer. *Peen* ramming involves the use of a rib-shaped edge to develop high impact pressures and is used principally to pack sand between narrow vertical walls and around the edges of the flask. *Butt* ramming is done with a broader-faced tool for more uniform compaction of the sand throughout the mold.

Jolting and Squeezing Use Mechanical Energy. Most production work and a large part of work done in small quantities is performed by use of molding machines whose principal duty is that of sand compaction. They are designed to compact sand by either *jolting* or *squeezing*, or both methods may be combined in a single machine.

Jolt compaction involves the lifting of the table carrying the mold and dropping it against a solid obstruction. With the sudden stop, inertia forces cause the sand particles to compress together. Jolt compaction tends to pack the sand more tightly near the parting surface. For this reason, it is usually not too satisfactory when used alone with patterns that are high and project close to the mold surface.

On the other hand, squeeze compaction, applied by pushing a squeeze plate against the outside of the sand, tends to pack the sand more tightly at the surface. The combination of jolting and squeezing is frequently used to take advantages of each method, although when both the cope and drag are being made on the same machine, it may be impossible to jolt the cope half (the second half constructed) without damage to the drag.

Sand Slinging Limited to Large Molds. Foundries that manufacture quantities of large castings often use sand *slingers* to fill and compact the sand in large floor molds. The sand is thrown with high velocity in a steady stream by a rotating impeller and is compacted by impact as it fills up in the mold. Figure 8-18 illustrates the common compaction methods.

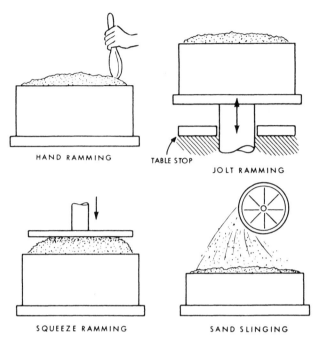

Figure 8-18
Common sand-compaction methods

CORES

Cores are bodies of mold material, usually in the form of inserts that exclude metal flow to form internal surfaces in a casting. The body is considered to be a core when made of green sand only if it extends through the cavity to form a hole in the casting. Green sand cores are formed in the pattern with the regular molding procedure.

Cores Need Strength for Handling. The vast majority of cores are made of dry sand and contain little or no clay. A nearly pure sand is combined with additives that burn out after pouring to promote collapsability and with binders to hold the particles together until after solidification takes place.

Final Core Properties Very Important. The properties needed in core sand are similar to those required for molding sand, with some taking on greater importance because of differences in the cores' position and use. Most cores are baked for drying and development of dry strength, but they must also have sufficient green strength to be handled before baking.

The *dry strength* of a finished core must be sufficient that it can withstand its own weight without sagging in the mold, and it must be strong enough that its own buoyancy, as liquid metal rises around it, will not cause it to break or shift.

Permeability is important with all molding sands but is especially so with core sand because cores are often almost completely surrounded by metal, and a relatively free passage is essential for the gases to escape through core prints or other small areas.

Collapsability is likewise important because of this metal enclosure. Ideally, a core should collapse immediately after metal solidification takes place. In addition to not interfering with shrinkage of the casting, it is important in many cases that cores collapse completely before final cooling so that they can be removed from inside castings in which they are almost totally enclosed. For example, cores used to form the channels in a hot-water radiator or the water openings in an internal combustion engine would be almost impossible to remove unless they lost their strength and became free sand grains. The casting metal must supply the heat for the final burning out of the additives and the binding material.

When a substantial portion of a core is enclosed in a casting, radiography is frequently used to determine whether or not the core shifted during casting, or to be certain that all the core material has been successfully removed after casting.

Chaplets. Very large or long slender cores that might give way under pressure of the flowing metal are sometimes given additional support by the use of *chaplets*. Chaplets are small metal supports with broad surfaced ends, usually made of the same metal as that to be poured, that can be set between the mold cavity and the core. Chaplets become part of the casting after they have served their function of supporting cores while the metal is liquid.

NDT may be necessary for castings requiring the use of chaplets. Not ony must the chaplets be chosen of suitable material to fuse with the base metal, but shrink cavities may form during the cooling, porosity may form from moisture condensation, and non-fusing may occur from too low a pouring temperature to melt the surface of the chaplet. Radiography of the finished casting can reveal discontinuities surrounding chaplet regions and can indicate whether the chaplets completely fused with the base metal.

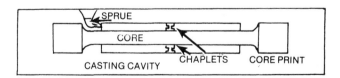

Figure 8-19
Slender core supported by chaplets to aid core location and prevent sagging of its own weight or springing, possibly floating, during pouring

GREEN SAND ADVANTAGES AND LIMITATIONS

Green Sand Process Extremely Flexible. For most metals and most sizes and shapes of castings, green sand molding is the most economical of all the molding processes. Green sand can be worked manually or mechanically and, because very little special equipment is necessary, can be easily and cheaply used for a great variety of products. The sand is reusable with only slight additions necessary to correct its composition. In terms of cost, the green sand process can be bested only when the quantity of like castings is large enough that reduced operational costs for some other processes will more than cover higher original investment or when the limitations of the green sand process prevent consistent meeting of required qualities.

Green Sand Not Universally Applicable. One of the limitations of green sand is its low strength in thin sections. It cannot be used satisfactorily for casting thin fins or long, thin projections. Green sand also tends to crush and shift under the weight of very heavy sections. This same weakness makes the casting of intricate shapes difficult also. The moisture present in green sand produces steam when contacted by hot metal. Inability of the steam and other gases to escape causes problems with some casting designs, and blowhole damage results. The dimensional accuracy of green sand castings is limited. Even with small castings, it is seldom that dimensions can be held closer together than ± 0.5 millimeter (0.02 inch); with large castings, ± 3 millimeters (1/8 inch) or greater tolerances are necessary.

DRY SAND MOLDS

Elimination of Moisture Reduces Casting Defects. Improvement in casting qualities can sometimes be obtained by use of *dry* sand molds. The molds are made of green sand modified to favor the dry properties and then dried in an oven. The absence of moisture eliminates the formation of water vapor and reduces the type of casting defects that are due to gas formation. The cost of heat, the time required for drying the mold, and the difficulty of handling heavy molds without damage make the process expensive compared to green sand molding, and it is used mostly when steam formation from the moisture present would be a serious problem.

Skin Drying — Substitute for Oven Drying. Most of the benefits of dry sand molds can be obtained by *skin drying* molds to depths from a fraction of an inch to an inch. With the mold open, the inside surfaces are subjected to heat from torches, radiant lamps, hot dry air, or electric heating elements to form a dry insulating skin around the mold cavity. Skin-dried molds can be stored only for short periods of time before pouring, since the water in the main body of the mold will redistribute itself and remoisturize the inside skin.

FLOOR AND PIT MOLDS

Large Molds Difficult to Handle. Although he number of extremely large castings is relatively small, molds must be constructed for one, five, ten, and occasionally, even as much as several hundred ton castings. Such molds cannot be moved about, and the high hydrostatic pressures established by high columns of liquid metal require special mold construction stronger than that used for small castings. *Floor* molds made in the pouring position are built in large flasks. The mold can be opened by lifting the cope with an overhead crane, but the cope flask usually must be constructed with special support bars to prevent the mold material from dropping free when it is lifted.

Drag of Pit Molds Below Floor Level. *Pit* molds use the four walls of a pit as a flask for the drag section. The cope may be an assembly of core sand or may be made in a large flask similar to that used for a floor mold. The mold material for these large sizes is usually loam, 50% sand and 50% clay, plus water. The mold structure is often strengthened by inserting bricks or other ceramic material as a large part of its substance.

SHELL MOLDS

Shell molding is a fairly recent development that, as far as casting is concerned, can be considered a precision process. Dimensions can be held within a few thousandths of an inch in many cases to eliminate or reduce machining that might be necessary otherwise and to decrease the overall cost of manufacturing. The cost of the process itself, however, is relatively high, and large quantities are necessary for economical operation.

Sand Bonded with Thermosetting Plastic. The mold is made by covering a heated metal pattern with sand that is mixed with small particles of a thermosetting plastic. The heat of the pattern causes the mixture to adhere and semicures the plastic for a short depth. The thin shell thus made is baked in place or stripped from the pattern, further cured by baking at 300° C and then cemented to its mating half to complete the mold proper. Because the shell is thin, approximately 3 millimeters, its resistance to springing apart is low; it may be necessary to back it up with loose sand or shot to take the pressures set up by filling with liquid metal. The sand particles are tightly held in the plastic bond. As erosion and metal penetration are minor problems, high quality surface finishes, in addition to good dimensional control, are obtained from shell molding.

METAL MOLD AND SPECIAL PROCESSES

Metal patterns and metal core boxes are used in connections with molding whenever the quantities manufactured justify the additional expense of the longer wearing patterns. The metal mold process refers not to the pattern equipment but to a reusable metal mold that is in itself a reverse pattern in which the casting is made directly.

Special Processes Receive Limited Use. In addition to the metal mold processes, there are special processes involving either single-use or reusable molds. Their use is limited to a comparatively small number of applications in which the processes, even though more costly, show distinct advantages over the more commonly used methods.

PERMANENT MOLD CASTING

Metal Molds Used Mostly for Low Melting Point Alloys. *Permanent* molds may be reused many times. The life will depend, to a large extent, upon the intricacy of the casting design and the temperature of the metal that is poured into the mold. Cast iron and steel are the most common materials with which the mold is made. Permanent mold casting is used most for the shaping of aluminum, copper, magnesium, and zinc alloys. Cast iron is occasionally poured in permanent molds that have much lower mold life because of the higher operating temperature. Satisfactory results require operation of the process with a uniform cycle time to maintain the operating temperature within a small range. Initial use of new molds often demands experimentation to determine the most suitable pouring and operating temperatures as well as to correct the position and size of the small vent grooves cut at the mold parting line to allow the escape of gases.

High Accuracies and Good Finishes. The cost of the molds, sometimes referred to as dies, and the operating mechanism by which they are opened and closed is high, but permanent mold casting has several advantages over sand casting for high quantity production. Dimensional tolerances are more consistent and can be held to approximately ±0.25 millimeter (0.1 inch). The higher conductance of heat through the metal mold causes a chilling action, producing finer grain structure and harder, stronger castings.

The minimum practical section thickness for permanent molding is about 3 millimeters (1/8 inch). The majority of castings are less than 30 centimeters (12 inches) in diameter and 10 kilograms (22 pounds) in weight. The process is used in the manufacture of automobile cylinder heads, automobile pistons, low horsepower engine connecting rods, and many other nonferrous alloy castings needed in large quantity.

DIE CASTING

Die casting differs from permanent mold casting in that pressure is applied to the liquid metal to cause it to flow rapidly and uniformly into the cavity of the mold, or die. The die is similar to that used for permanent molding. It is made of metal, again usually cast iron or steel; has parting lines along which it can be opened for extraction of the casting; and is constructed with small draft angles on the walls to reduce the work of extraction and extend the life of the die. Vents, in the form of grooves or small holes, also are present to permit the escape of air as metal fills the die.

Hot Chamber Die Casting. The machines in which the dies are used, however, are quite different because, in addition to closing and opening the die parts, they must supply liquid metal under pressure to fill the cavity. The *hot chamber* die-casting machine, as shown in Figure 8-20, keeps metal melted in a chamber through which a piston moves into a cylinder to build up pressure forcing the metal into the die.

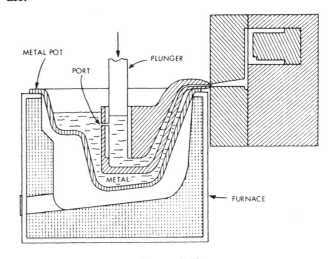

Figure 8-20
Hot chamber die casting

Machines Limited to Low Pressures. Because the piston and the portions subjected to pressure are heated to the melting temperature of the casting metal, hot chamber machines are restricted to lower pressures than those with lower operating temperatures. Although it is a high speed, low cost process, the low pressures do not produce the high density, high quality castings often desired. In addition, iron absorbed by aluminum in a hot chamber machine would be detrimental to its properties. Pressures as high as 14 MPa (2,000 psi) are used in the hot chamber process to force fill the mold.

Cold Chamber Die Casting. With *cold chamber* equipment, as shown in Figure 8-21, molten metal is poured into the shot chamber, and the piston advances to force the metal into the die. Aluminum, copper, and magnesium alloys are die cast by this method with liquid pressures as high as 210 MPa (30,000 psi).

Casting Quality High. Sections as thin as 0.4 millimeter (1/64 inch) with tolerances as small as

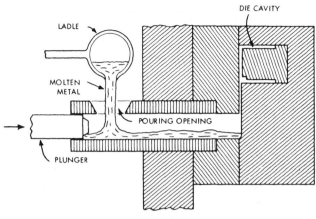

Figure 8-21
Cold chamber die casting

0.05 millimeter (0.002 inch) can be cast with very good surface finish by this pressure process. The material properties are likely to be high because the pressure improves the metal density (fewer voids), and fast cooling by the metal molds produces good strength properties. Other than high initial cost, the principal limiting feature of die casting is that it cannot be used for the very high strength materials. However, low temperature alloys are continually being developed, and with their improvement, die casting is being used more and more.

INVESTMENT CASTING

The Working Pattern Destroyed During Investment Casting. Investment casting (Figure 8-22) is also known as precision casting and as the *lost wax* process. The process has been used in dentistry for many years. A new wax pattern is needed for every piece cast. For single-piece casting, the wax pattern may be made directly by impressions as in dentistry, by molding or sculpturing as in the making of statuary, or by any method that will shape the wax to the form desired in the casting. Shrinkage allowances must be made for the wax, if it is done hot, and for the contraction of the metal that will be poured in the cavity formed by the wax. Reentrant angles in the casting are possible because the wax will not be removed from the cavity in solid form. Variations of this process involve the use of frozen mercury or low melting point thermoplastics for the pattern.

Duplicate Parts Start with a Master Pattern. Multiple production requires starting with a master pattern about which a metal die is made. The metal die can be used for making any number of wax patterns. A gating system must be part of the wax pattern and may be produced in the metal die or attached after removal from the die. When complete, the wax pattern is dipped in a slurry of fine refractory material and then encased in the investment material (plaster of paris or mixtures of ceramic materials with high refractory properties). The wax is then removed from the mold by heating to liquify the wax and cause it to run out to be reclaimed. Investment molds are pre-

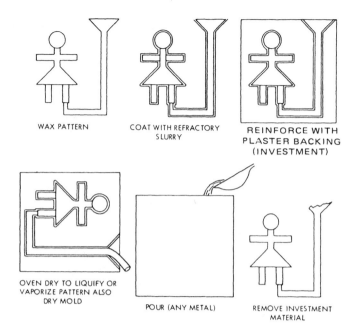

Figure 8-22
Steps for investment casting

heated to suitable temperatures for pouring, usually between 600° C and 1,100° C, depending upon the metal that is to fill the mold. After pouring and solidification, the investment is broken away to free the casting for removal of the gating system and final cleaning.

Process Limited to Small Castings. Investment casting is limited to small castings, usually not over 2 kilograms (4.4 pounds) in weight. The principal advantage of the process is its ability to produce intricate castings with close dimensional tolerances. High melting temperature materials that are difficult to cast by other methods can be cast this way because the investment material of the mold can be chosen for refractory properties that can withstand these higher temperatures. In many cases, pressure is applied to the molten metal to improve flow and densities so that very thin sections can be poured by this method.

High Quality at High Cost. It can easily be realized, by examination of the procedures that must be followed for investment molding and casting, that the costs of this process are high. Accuracy of the finished product, which may eliminate or reduce machining problems, can more than compensate for the high casting cost with some materials and for some applications.

A number of important parts, some of new or exotic materials, are presently manufactured by investment casting. Many of these, such as high strength alloy turbine buckets for gas turbines, require NDT inspection by radiographic and penetrant methods to insure that only parts of high quality get into service.

PLASTER MOLD CASTING

Molds made of plaster of paris with additives, such as talc, asbestos, silica flour, sand, and other materials to vary the mold properties, are used only for casting nonferrous metals. Plaster molds will produce good quality finish and good dimensional accuracy as well as intricate detail. The procedure is similar to that used in dry sand molding. The plaster material must be given time to solidify after being coated over the pattern and is completely oven dried after removal before it is poured.

Casting Cools Slowly. The dry mold is a good insulator, which serves both as an advantage and as a disadvantage. The insulating property permits lower pouring rates with less superheat in the liquid metal. These contribute to less shrinkage, less gas entrapment from turbulence, and greater opportunity for evolved gases to escape from the metal before solidification. On the other hand, because of slow cooling, plaster molds should not be used for applications in which large grain growth is a serious problem.

CENTRIFUGAL CASTING

Several procedures (Figure 8-23) are classed as centrifugal casting. All of the procedures make use of a rotating mold to develop centrifugal force acting on the metal to improve its density toward the outside of the mold.

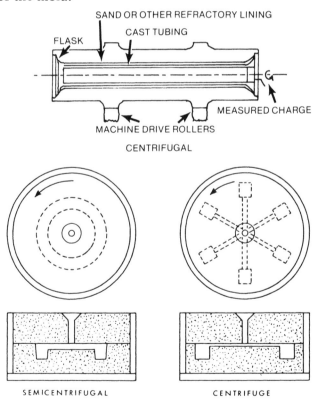

Figure 8-23
Centrifugal casting

True Centrifugal Casting—Hollow Product. The true centrifugal casting process shapes the outside of the product with a mold but depends upon centrifugal force developed by spinning the mold to form the inside surface by forcing the liquid metal to assume a cylindrical shape symmetric about the mold axis. At one time the principal product was cast iron sewer pipe, but present day uses of centrifugal castings include shafts for large turbines, propeller shafts for ships, and high pressure piping. Because of the critical nature of some applications NDT may be necessary to check the wall thickness and quality of the product material. The columnar grain structure may produce problems in applying nondestructive tests.

Semicentrifugal Casting — Solid Product. A similar process, which may be termed semicentrifugal casting, consists of revolving a symmetric mold about the axis of the mold's cavity and pouring that cavity full. The density of a casting made in this way will vary, with dense, strong metal around the outside and more porous, weaker metal at the center. The variation in density is not great, but the fast filling of the external portion of the mold cavity produces particularly sound metal. Wheels, pulleys, gear blanks, and other shapes of this kind may be made in this way to obtain maximum metal properties near the outside periphery.

Centrifuge Casting — Multiple Product. A third type of casting using centrifugal force can be termed centrifuge casting. In this process, a number of equally spaced mold cavities are arranged in a circle about a central pouring sprue. The mold may be single or stacked with a number of layers arranged vertically about a common sprue. The mold is revolved with the sprue as an axis and when poured, centrifugal force helps the normal hydrostatic pressure force metal into the spinning mold cavities. Gases tend to be forced out of the metal, which improves metal quality.

CONTINUOUS CASTING

Although only a small tonnage of castings are produced by continuous casting, it is possible to produce two-dimensional shapes in an elongated bar by drawing solidified metal from a water-cooled mold.

Special Equipment and Skills Required. As shown schematically in Figure 8-24, molten metal enters one end of the mold, and solid metal is drawn from the other. Control of the mold temperature and the speed of drawing is essential for satisfactory results.

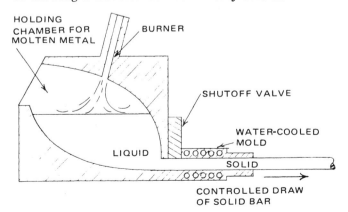

Figure 8-24
Schematic diagram of continuous casting process

Good Quality Castings Possible. Exclusion of contact with oxygen, while molten and during solidification produces high quality metal. Gears and other shapes in small sizes can be cast in bar form and later sliced into multiple parts.

An automotive manufacturer makes use of the concept as a salvage procedure for saving bar ends of alloy steel. The waste material is melted and drawn through the mold in bar form. Subsequently, the bars are cut into billets that are suitable for processing into various automotive parts.

MELTING EQUIPMENT

The volume of metal needed at any one time for casting varies from a few pounds for simple castings to several tons in a batch type operation with a continuous supply, usually of iron, being required by some large production foundries. The quantity of available metal can be varied by the size and type of melting equipment as well as the number of units in operation. The required melting temperature which varies from about 200°C (390°F) for lead and bismuth to as high as 1540°C (2400°F) for some steels also influences the type of melting equipment that will serve best.

CUPOLA

A considerable amount of cast iron is melted in a special chimney-like furnace called a cupola. It is similar to a blast furnace (described in Chapter 5) used for refining iron ore. The cupola (Figure 8-25) is charged through a door above the melting zone with layers of coke, iron, and limestone and may be operated continuously by taking off melted iron as it accumulates in the well at the bottom.

CRUCIBLE FURNACES

Melting of small quantities (1 to 100 pounds) of non-ferrous materials for small volume work is often performed in lift out crucibles constructed of graphite, silicon carbide, or other refractory material. Gas or oil is combined with an air blast around the crucible to produce the melting heat. Unless a cover is placed on the crucible, the melt is exposed to products of combustion and is susceptible to contamination that may reduce the quality of the final castings. This is true of all the natural fuel fired furnaces.

POT FURNACES

Quantities of non-ferrous materials to several hundred pounds may be melted in pot furnaces that contain a permanently placed crucible. Metal is ladled directly from the crucible, or in the larger size equipment, the entire furnace is tilted to pour the molten metal into a transporting ladle.

REVERBERATORY FURNACES

Some of the largest foundries melt non-ferrous metals in reverberatory furnaces that play a gas-air or oil-air flame through nozzles in the side walls of a brick structure, directly on the surface of the charged material. Gas absorption from products of combustion is high but the large capacity available and high melting rate provide economics that help compensate for this fault. Smaller tilting type reverberating furnaces are also available for fast melting of smaller quantities of metal.

ELECTRIC ARC FURNACES

The electric arc provides a high intensity heat source that can be used to melt any metal that is commonly cast. Since there are no products of combustion and oxygen can be largely excluded from contact with the melt, quality of the resulting cast metal is usually high.

The arc may be *direct* (between an electrode and the charged metal) or *indirect* (between two electrodes above the charge).

INDUCTION FURNACES

Induction furnaces melt materials with the heat dissipated from eddy currents. Coils built into the furnace walls set up a high frequency alternating magnetic field which in turn causes internal eddy currents that heat the charge to its melting point. Rapid heating and high quality resulting from the absence of combustion products help offset the high cost of the equipment and power consumed.

FOUNDRY MECHANIZATION

The preceding pages briefly describe the most common foundry techniques for producing castings. Most are performed largely by manual effort, resulting in relatively slow production. However, at any time the production quantities justify the needed expenditure for equipment, these same techniques are subject to almost complete mechanization resulting in higher production rates and improved consistency.

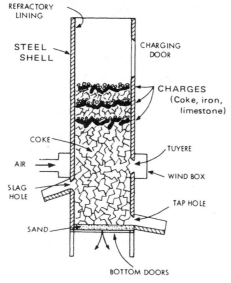

Figure 8-25
Cupola

The Welding Process 9

Welding is a *joining* procedure in which shape changes are only minor in character and local in effect. Welding may be defined as "the permanent union of metallic surfaces by establishing atom to atom bonds between the surfaces." In practice, some distinction is usually made between true welding and brazing and soldering. In true welding, the filler material has a composition similar to that of the base metal(s). In brazing and soldering, however, the filler is a metal with a lower melting point than the base metal(s). Adhesive joining, which is sometimes performed as a true welding process with certain plastics, usually makes use of organic adhesives, often containing plastic filters and inorganic solvents that fuse the surfaces of the plastic and adhesive together. With some plastics, a sound plastic/plastic joint can be formed by only introducing a volatile solvent into the joint which "melts" the plastic/plastic interface, essentially welding the parts together. Adhesive joining will be discussed in Chapter 15, Miscellaneous Processes.

Development of Welding Relatively Recent. Welding is both an ancient and a new art. Evidence indicates that prehistoric man, finding native metals in small pieces and being unable to melt them, built up larger pieces by heating and welding by hammering or forging. On the other hand, arc welding was first used in 1880 and oxyacetylene in 1895. Even after these developments, welding remained a minor process, used primarily as a last resort in maintenance and repair, until about 1930. After this date, the increased knowledge of metallurgy and testing and the development of improved techniques led to increased confidence and use, so that today welding may be considered a basic shape-producing method in direct competition with forging, casting, machining, and the other important processes.

Versatility Provides Many Applications. While it is true that welding itself does not change the shape of the individual components, the finished *weldment*, or assembly of parts, constitutes a unified structure that functionally has the properties of a solid part. In some cases, particularly with spot welding, welding is purely an assembly procedure and competes with mechanical fastening, such as riveting or bolting. In other cases, the goal in welding is to provide a joint that has the same structure, strength, and other properties as the base metal so that the weld area itself would be undetectable. This goal is approached in producing some pipe and high pressure vessels but usually requires elaborate precautions to prevent contamination, heat treatment of the entire weldment after welding, and thorough testing, usually by radiography. In most cases, these procedures would not be practical or economical; consequently, some reinforcement of the welded area is provided by designing with reinforcing plates or gussets.

Often Replaces Bolting and Riveting. With the exception of some of the special purpose techniques in other areas, welding is in a greater period of growth than any of the other manufacturing procedures. Welding has largely replaced riveting and bolting in structural steel work for bridges and buildings. In the manufacture of automobiles and home appliances from sheet metal, most of the joining of large shapes is by welding, and in many cases these welds are not even apparent in the finished product. A typical automobile, for example, has over 4,500 spot welds in addition to other welding.

BONDS

NATURE OF BONDING

Atomic Bonding Essential. Most welding definitions include some reference to heat and pressure, and in practice most welding processes do make use of heat or pressure or both. However, neither of these is theoretically necessary. If two perfectly matched clean surfaces are brought together within suitable atomic spacing, atomic bonds will automatically be established between the surfaces, and the surfaces will, in fact, be welded. The essential features are not so easy to realize, however.

Atomic Cleanliness and Closeness. Atomic cleanliness requires that atoms exposed on the surfaces actually be the atoms of the materials to be joined. Even if this condition is set up on a surface, exposure to the atmosphere results in almost immediate formation of oxide or sulfide films on most metals. Atomic closeness requires that the distances between atoms brought into contact be that at which atoms are normally spaced in the crystalline structure of a metal. Normally, when two surfaces are brought into contact, this condition will occur only at a number of points because surfaces of even the best quality have a finite roughness of a much larger order than atomic distances.

Melting Common but Not Essential. Various means may be used to establish these two essential conditions of atomic cleanliness and closeness. Cleanliness may be established by chemical cleaning (fluxing), providing the products of the cleaning operation may be removed from the surface; by melting the surface area so that the surface films float to the surface of the molten material; or by fragmentation as a result of plastic deformation of the base metal. Atomic closeness may be established by filling with a liquid metal, as in brazing and soldering, without actually melting the joined metals; by elastically or plastically deforming the surfaces until contact is established; or by actually destroying the surfaces by melting and allowing molten base metal or melted filler material to resolidify in contact with the unmelted base metal.

Welding may be accomplished as a result of any combination of conditions that establishes the two essential elements of atomic cleanliness and atomic closeness.

FUSION BONDING

Most important welding processes, particularly those in which high strengths are a principal goal, make use of *fusion* bonds in which the surfaces of the pieces to be joined (*parent* or *base* metal) are completely melted, as shown in Figure 9-1. Liquid metal then flows together to form the union, and cleanliness is established as the impurities float to the surface. No pressure is necessary, and the parts to be joined need only be located and held in proper relationship to each other.

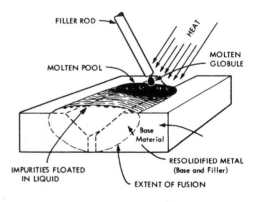

Figure 9-1
Fusion bond

Metallurgical Effects Like Casting. The resolidification of the metal results in a localized casting for which the unmelted base metal serves as a mold. It can then be expected that the same metallurgical changes and effects, such as grain-size variation and shrinkage, that occur in casting will occur in fusion welding. It is also implied that simply heating an entire structure that is to be fusion bonded would not be satisfactory because the entire structure would reach the melting temperature at the same time. The heat must be supplied locally to the area to be melted, and the rate of heat input must be great enough to prevent overheating of the adjacent areas. This requirement leads to some difficulties in welding aluminum, copper, and other metals having very high thermal conductivities. Hence, NDT for weldments is similar to that for castings. The same kinds of defects are likely to be found and similar NDT methods may be effective. In most cases there is no advantage in inspecting the entire weldment because the weld defects will be concentrated in the weld itself or in the heat affected zone of the weld.

Filler Sometimes Added. In fusion welding, at least the surface of the parts being welded is always melted, and this amount of molten metal may be sufficient to form the weld. In the more used fusion-welding processes, however, additional molten metal (*filler*) is supplied, usually by continuously melting a rod or wire. The use of filler is nearly always necessary in welding sheet and structural shapes more than 3 millimeters (1/8 inch) in thickness and, in many cases, permits more freedom in joint design by making possible the filling and buildup of gaps and cavities.

Welded Joint Strength. The strength of fusion-welded joints will depend on the composition and metallurgical structure of the filler material and base metal, on any structural changes that take place in heated areas of the base metal adjacent to the weld, on the perfection with which the desired geometry of the weld is established, on residual stresses built up as a result of the differential heating and cooling, and on the presence or absence of impurities in the weld. It is at least theoretically possible to produce 100% efficiency in a fusion weld as compared to unwelded base metal.

PRESSURE BONDING

Heat Aids Cleanliness and Closeness. The term *pressure bonding* is somewhat misleading in that some heating is involved in the processes called pressure bonding, or pressure welding. As will be discussed later, pressure alone may be sufficient to form a bond, but heat is used for two principal reasons. The close union required is established by plastic flow, as indicated in Figure 9-2, and, in general, metals become more plastic, and strengths are lower as the temperature is raised. Pressure and flow cause some fragmentation of the oxides on the surface because most are quite brittle and cannot maintain a continuous film as the metal flows plastically.

Of even greater importance are the two effects heat has on this oxide layer, which must be removed or dispersed before bonding can be effective. First, the fragments tend to assume spherical shapes as their total energy is raised. With spherical shapes, they disrupt a lower percentage of the surface areas to be joined, allowing greater contact between exposed base materials. This same type of spheroidization accounts for the malleabilizing of cast iron and for the effect on the cementite particles in the prolonged heating of high carbon steels. The second heat effect is that the solubility of oxygen in the base metals is raised with increased temperature, and some dissociation of the oxides occurs with the oxygen being diffused into the base metal.

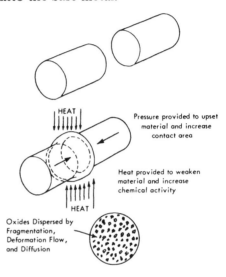

Figure 9-2
Pressure bond

Overall Joint Efficiency High. While a small amount of fusion of the base metal may occur in some pressure bonds, it is incidental. No pronounced solidification shrinkage occurs as it does with fusion welds. Consequently, distortion after welding is usually very slight. The efficiency of pressure bonds, based on the original area, may be as high as 95%. Even though there are some inclusions in the weld area that lower unit strength, pressure-welded joints may actually be stronger than the original cross section as a result of the enlargement that occurs with plastic flow. This is especially true in butt-welding procedures as used in the manufacture of some chain links and fittings.

FLOW BONDING

Base Material Not Melted. When a filler material of different composition and lower melting temperature than the base metal is used, the mechanism is described as *flow* bonding (Figure 9-3). While some fusion of the base metal may occur, it is not essential to the process and is usually undesirable. The closeness necessary for bonding is established by the molten filler metal conforming to the surface of the base metal. The required cleanliness is produced by use of fluxes, ordinarily metal halides or borax, which dissolve the surface oxides and float them out of the joint.

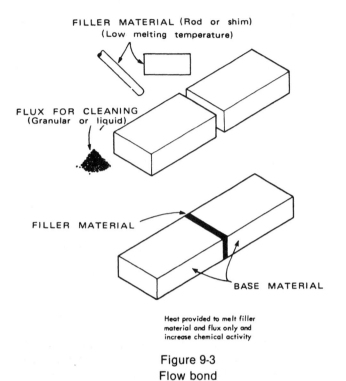

Figure 9-3
Flow bond

Joint Defined by Temperature and Spacing. Three different operations using flow bonds have been named: braze welding, brazing, and soldering. In *braze welding*, the filler material is a metal or alloy having a melting point above 425° C (800° F) and a composition significantly different from the base metal. In practice, the commonest alloys used as filler are copper or silver based. Occasionally, pure copper is used for braze welding steel. The filler is usually in rod form, and the procedures are similar to those employed in some fusion welding except that only the filler material is melted. Fluxes are heated on the joint surfaces for cleaning. Braze welding is used mainly for joining and repairing cast iron and is being replaced by fusion welding in many cases. The joint strength is limited to that of the filler material in cast form.

Brazing. The word *brazing*, when used alone, designates the use of filler materials similar to those used in braze welding but applied to a close-fitting joint by preplacement or by capillary action. Filler material may be rod, wire, foil, slug, or powder, and fluxes similar to those used in braze welding are necessary. Heat may be furnished by torch, furnace, or induction, and, in production quantities, by dipping in molten salts, which may also provide the fluxing action.

Brazing quality depends upon the proper combination of base metal properties and joint preparation, filler metal properties, brazing temperature, and time at temperature. All factors in this combination are significant to provide for melting the braze filler metal, causing it to flow, fill, wet the joint, and diffuse into the surface layers of the atomic structure of the base metal.

Joint Thickness Critical to Strength. Figure 9-14 shows the importance of thickness to the strength of a brazed joint. The low strength of very thin joints is due to the formation of "capillary dams" caused by uneven surfaces that prevent complete filling. This fault can be overcome to some extent by use of special techniques, such as application of ultrasonic vibration while brazing. The fact that the strength of the joint can be higher than that of cast filler is due to the differences in modulus of elasticity between the filler and the base material. The filler metal is prevented from yielding by the more rigid base metal; the result is high shear stresses normal to the direction of the load in the filler material. These shear stresses generate tensile stresses in such direction that when they are combined vectorially with the direct tensile stresses caused by the load, a lower stress value is produced on the plane normal to the load than would occur in a homogeneous material. When the joint becomes thicker, there is less restraint in the center of the filler layer, the shear stresses are lower, and their effect in compensating for direct load stresses is reduced.

Brazing is frequently used to join parts together, particularly when one or more of those parts would be subject to changes from exposure to high temperature (above that needed for brazing). If the joint strength is critical or if leakage is a factor, NDT might well be used to establish that the necessary joint quality exists. The worst possible fault (assuming the braze itself is complete) would be wide spacing, either total or partial, due to poor preparation, angular geometry, or wide positioning. Sloppy fit-up of joints can also cause the molten braze filler to fail to completely fill the joint. Obviously, such conditions also produce joints of very low strength. Porosity and inclusions are other possible defects. In critical applications, either ultrasonic or radiographic tests may be used to check the joint quality. Radiography readily reveals unfilled joints and porosity. However, unwetted and undiffused joints can seldom be revealed by radiography. If such conditions are suggested, ultrasonic techniques should be used; ultrasonic transmission characteristics through a properly wetted and diffused joint are significantly different from those through an inadequate joint.

Soldering. The third type of flow bonding, *soldering*, actually includes application similar to both braze welding and brazing. The essential difference is

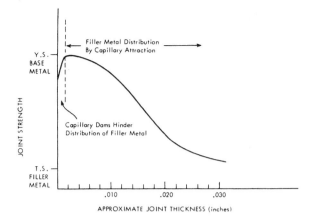

Figure 9-4
Strength of brazed joints

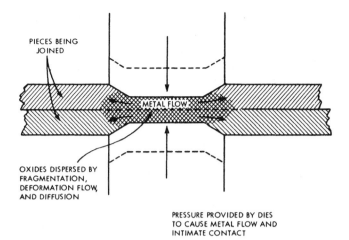

Figure 9-5
Cold bond

in the melting temperature of the filler metal, which for soldering is below 425° C (800° F). The most important materials in this class are lead-tin alloys with melting points from 185° C (361° F) to slightly above 315° C (600° F). The mechanical strengths of soldered joints, particularly built-up joints of the braze-weld type, are low, and the greatest use for soldering is for providing fluid tightness, for electrical connections, and for sheet metal joint filling in automotive assembly work.

COLD BONDING

Heat Not Essential to Bonding. In fusion, pressure, and flow bonding, heat is used to help establish the closeness and cleanliness necessary, but heat, as such, is not essential for proper bonding between metallic surfaces. With greater loads than used in pressure bonding, plastic flow of the required order for fragmentation of surface impurities can be established in ductile materials at room temperature. If two fresh surfaces of lead are twisted together, a weld is made with a strength approaching that of the base metal, and any metal may be made to weld to some degree by wiping two surfaces together at sufficiently high normal pressure. However, the results would be inconsistent. The practical application of cold bonding depends on inducing deformation parallel to the interface while it is subjected to high normal pressure.

Contact Area Increased. In practice, welds are made by squeezing the metal between two punch faces that cause metal flow normal to the direction of load (Figure 9-5). As the area of contact is increased, the brittle surface oxides fragment and cover a smaller percentage of the area, exposing clean metal to metal contacts. The greatest success so far has been with copper and aluminum base metals.

WELDING METALLURGY

Welding Introduces Complex Problems. The final properties of a welded or brazed joint are influenced by many factors. Some of the metal is actually melted in most cases, and welded parts are subject to deformation and high shrinkage on cooling. The metallurgical changes in a weldment may include all that take place under any kind of processing, including melting, alloying, solidification, casting, hot and cold working, recrystallization, and heat treating. In the case of welding, most changes are intensified because of the high thermal gradients developed and the fast rates of heating and cooling encountered. These side effects are often overlooked or neglected because the principal objective of the welding procedure is the joining of material.

The conditions under which most welding is performed are far from ideal. Tremendous energy inputs, especially in fusion welding, may lead to localized overheating to the point of vaporization. Exposure of high temperature and chemically active materials to atmospheres difficult to control leads to the formation of undesirable compounds. Most gases are highly soluble in molten metals but have decreasing ability to stay in solution as temperatures lower, leading to problems of gas entrapment.

COMPOSITION EFFECTS

Dissolved and Entrapped Gases. The conditions existing in the weld area are frequently conducive to significant changes in the composition of either the base or the filler metal. The rapid solidification rates may lead to segregation of some elements, particularly coring-type segregation as may occur in casting some brasses. Gas may enter the molten metal not only by solution but also as a result of agitation that occurs with many fusion-welding processes. These entrapped gases can form voids or brittle compounds within the structure of the metal. One of the most serious conditions is the embrittlement resulting from hydrogen trapped in steel. With the rapid solidification, slag and oxides may not have time to float and may be trapped beneath the surface to appear as solid inclusions in the completed weld.

Uniform Structure Possible. Ideally, it is possible to produce a fully homogeneous material without defects in which it would be difficult if not impossible to detect the welded metal. In practical applications, this situation can be approached but often requires post treatment to produce completely uniform structure and properties. Because the theoretically perfect joint is almost impossible to accomplish, inspection by NDT is frequently necessary to determine the degree of quality. It should be pointed out also that in those cases where the quality of a process is dependent to a large extent on the skill and care of an operator, quality of the work is likely to be higher if inspection is to be performed or even if the chance of inspection exists.

Filler May Reduce Problems. Fillers of composition different from that of the base metal are often used to compensate for welding faults that might otherwise be expected. The attempt is not usually to use a filler that will exactly compensate for the losses of the welding process but rather overcompensate for improvement of certain properties. Thus, high nickel filler may be used in welding cast iron to control grain growth and give ductility to the weld area, and stainless steel filler may be used with higher alloy content than the base material to insure adequate corrosion resistance. Brazing and soldering alloys are used principally to avoid high temperature effects in the base metal.

The amount of alloying that occurs between base metals and filler metals of different composition depends on several factors, but chiefly on the actual metals involved. Alloying is not essential to true bonding, but at the high temperatures reached, diffusion proceeds at a high rate, and for some metals alloying will occur for some distance in both directions away from the original interface. Soft solders in particular may produce brittle intermetallic compounds that reduce ductility and lower strength.

EFFECTS ON GRAIN SIZE AND STRUCTURE

Cooling Rates Higher Than in Casting. Grain formation in fusion welds can best be understood by remembering that a fusion weld is a casting, and all the effects present in casting will be duplicated. However, the mold wall is not fixed, and the solidification and cooling rates are faster than normally occur in casting (Figure 9-6). Fusion welds are subject to solidification and cooling shrinkage, as shown in Figure 9-7. The grain-size effects are not confined to the molten metal, because temperature high enough to result in annealing, allotropic transformation, and recrystallization extends for some distance into the base metal, as shown in Figure 9-8. The fused material is cooled rapidly by the high thermal conductivity of the surrounding metal, and small grain size results.

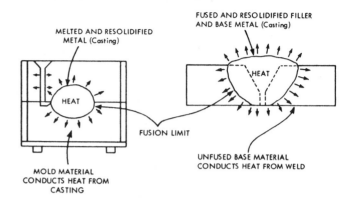

Figure 9-6
Comparison of fusion weld with casting

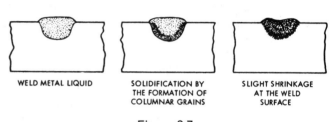

Figure 9-7
Solidification of a bead weld

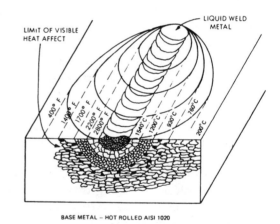

Figure 9-8
Grain structure in a fusion weld

Heat Affects Base Material. The zones indicated in the drawing do not have sharp dividing lines and represent only typical results. The results can vary from those shown, depending on the shape and size of the parts, the initial temperature of the base material, the rate of heat input, and the alloy content. In any case, for steels, an area immediately surrounding the molten metal will be heated above the transformation temperature, and some degree of austenitization can occur. Final results will depend on the time at temperature and the cooling rates, which cannot always be accurately predicted. Grain growth can proceed, and, for the metal heated near its melting temperature, the final grain size can be large. The metal heated only slightly above the transformation temperature is effectively normalized and will have a small final grain size, which can be smaller than that of the unheated base metal. Any heat-treat or cold-work hardening that existed in the area heated below the transformation temperature will be subject to tempering or recrystallization, depending on the actual temperature reached and the preweld condition.

When ultrasonic inspection is being performed on a weldment, it is important to recognize that the abrupt change in grain size can often be detected. The ultrasonic signal reflected from this heat-affected zone may be misinterpreted in some cases as being lack of fusion or a variety of other discontinuities, depending on location.

Multiple Cooling Rates. Again, depending on cooling rates induced and compositions involved, for the metal heated above the transformation temperature, the cooling may be equivalent to that required for annealing, normalizing, or actually quenching to martensite, provided enough carbon is present. Some of the latter nearly always occurs in unpreheated carbon steel weldments and, when combined with the uneven shrinkage that may be present, can result in brittle structures subject to cracking. Alloy rods or rods of different carbon content may be used for controlling some of the possible defects. Low carbon filler material is often used in welding higher carbon steels to avoid the formation of excessive amounts of martensite. In the fusion zone where cooling rates are high, the composition would be near the composition of the filler material. Even with rapid cooling, the structure would consist mainly of ferrite with sufficient ductility to shrink without cracking.

Structure Varies with Cooling Rate. In the base material adjacent to the liquid metal, the cooling rate would be somewhat less but still sufficiently rapid to form fine pearlite and some martensite. It must be remembered that grain size and structure are two different considerations; in this region, grain size will be large because of the long time at high temperature, but structure will be fine because of the rapid cooling. At a greater distance from the molten zone but still within the area raised above the transformation temperature, the cooling rate will be nearer that usual with normalizing, and the resulting structure will be medium to coarse pearlite.

Preheating Lowers Cooling Rate. The cooling rate of the weld and the entire weld area is changed by preheating the base metal surrounding the area to be welded. At any given point in the weld area, the cooling rate will be reduced because of the reduced thermal gradient established. Average grain size will be larger because of the longer times at high temperature, but structures will be softer because of the reduced cooling rates.

Effects in Pressure Welding Reduced. Effects similar to those of fusion welding will be observed in pressure welding. With lower temperatures, and frequently higher thermal gradient, the heat-affected zone will be smaller. Shrinkage problems are reduced because of little or no fusion and more uniformly welded cross sections.

EFFECTS OF WELDING ON PROPERTIES

Post treatment Sometimes Is Valuable. In an ideal weld, the composition of the weld zone could be made like that of the base metal and, with proper heat treatment, the strength of the final weldment would be unaffected in any way by the presence of the weld. In most practical situations, compositions cannot be kept exactly the same, and heat treatment sufficient to establish completely uniform structures would be uneconomical, if not impossible. The result is that the strength of most welds is different from that of the base metal. With no heat treatment of welded steel, the strength and hardness will vary from that of annealed to that of quenched material. Ductility will vary inversely with the strength. Many weldments are at least normalized to obtain more uniform properties and to relieve stresses.

Design Consideration Essential. The possible presence of discontinuities and inclusions in a weld may lead to reduced strengths for which consideration must be given in weld design. The designer must either gamble on weld quality, require special inspection procedures to determine weld quality with possible rewelding of some structures, or overdesign welded joints on the basis of lowest expected strengths.

Changes May Adversely Affect Corrosion Resistance. Corrosion resistance of many welded metals is likely to be affected adversely. As already pointed out, composition and structural changes accompany the usual conditions required to produce a weld. High temperatures lead to diffusion and precipitation effects that change the chemical characteristics of the metal. Some stainless steels are subject to the formation of chromium carbide during welding and may lose much of their corrosion-resistant qualities without proper subsequent heat treatment. Even with protective procedures, such as inert gas shielding or slag coverings, discoloration and surface oxidation occur in the heat-affected zone. Materials under high stress are subject to increased corrosion, and welds are prone to highly stressed areas unless special treatment is used for their removal.

DISTORTIONS AND STRESSES

A homogeneous unrestricted body may be heated to any temperature below its melting point without shape change. A volumetric expansion will occur with heating, but if this expansion occurs uniformly, no stresses will be introduced. As the body is cooled, the process reverses, and the final result will be the original unstrained state.

Restraints Create Stresses and Distortion. With restraint either on heating or cooling or with heating or cooling of localized areas at a more rapid rate than others (self-restraint), the picture will be changed. Many welds have a vee cross section, and the molten and heated areas will have a related shape. Furthermore, the heat input and higher temperatures occur on the open side of the vee. Figure 9-9 illustrates the result of cooling on this cross section for various weldments joined with vee welds. The greater shrinkage occurring on the wide side of the vee leads to angular distortion as shown. The effect is amplified by multipass welds in which a number of weld deposits are made along the length of a single vee. Each pass contributes to the distortion with the deposits from previous passes serving as a fulcrum for increased angular movement.

While a vee weld will always tend to distort angularly in the manner shown, the lateral distortion between members of a weldment may vary in direction and amount, depending on the size of the members compared to the weld, the number of passes made, the rate of heat input, and the speed of welding (Figure 9-10). As the weld proceeds along the groove, the heating of base metal along the edge of the groove but ahead of the acutal weld leads to a spreading of the plates. On the other hand, the

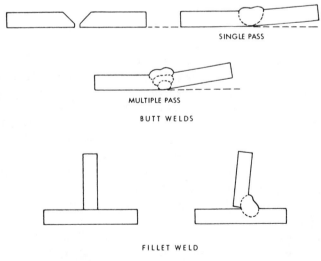

Figure 9-9
Angular distortion

shrinkage accompanying the solidification and cooling of the completed weld tends to pull the plates together.

All Welds Create Residual Stresses. Practical weldments never have absolute restraint or absolute freedom, and the actual degree of restraint and tempera-

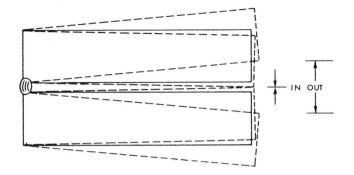

Figure 9-10
Lateral distortion

ture difference cannot always be predicted or measured. However, some degree of restraint always exists, at least in the parent metal adjacent to the weld zone, even for members that as a whole are free. It can be safely stated that any fusion weld will contain some residual stresses when completed and cooled to room temperature. These stresses will be both tensile and compressive because a balance must exist for the member to be in equilibrium.

Stresses and Distortion Are Associated. Some results are indicated in Figure 9-11. For a weld along the edge of a plate, the longitudinal shrinkage will cause curvature as indicated. Although the plate has no external restraint, it will be subject to stresses similar to those resulting from external loading that would cause equivalent curvature. In the case of the weldment, however, there will be two neutral axes with both edges in tension and the center under compression.

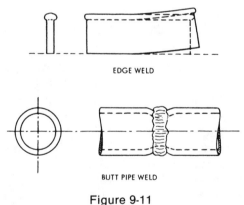

Figure 9-11
Longitudinal distortion

For a circular weld around a pipe, similar self-restraint exists. The shrinkage along the length of the weld results in a reduction in diameter that is resisted by the solid pipe adjacent to the weld. The result would be high tensile stresses in the weld and high compressive stresses in the pipe on both sides of the weld.

Even when the welded members have no external restraint or apparent gross distortion, high residual stresses can exist. Figure 9-12 indicates the kind of stress distribution to be expected from a longitudinal butt weld between two plates.

Stresses Reduced by Postheating. The most widely accepted method of reducing residual stresses in the weldments is based on two facts: (1) no stresses higher than the yield stress can exist in a material at any given temperature, and (2) if an entire unrestrained body is cooled uniformly from any given temperature, no increase in stress will occur. If a weldment is heated to an elevated temperature, yielding will occur and the stresses reduced. As the temperature is reduced, the entire weldment will shrink, but no new stresses will be introduced. Residual stresses cannot be completely eliminated by this method but, as the figure shows, the yield strength at elevated temperatures is quite low.

The stabilization of stresses in a weldment requires that the entire weldment reaches a uniform temperature and that all distortions permitted by its restraints take place. Time is required for each to happen and even after stabilization some residual stresses may be very near critical levels. It is sometimes important

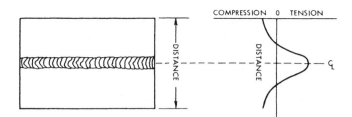

Figure 9-12
Longitudinal stress in a butt weld

that final inspection be delayed hours, or longer, to be certain that post-cracking will not occur shortly after an inspection has been made.

Grain Uniformity Requires Transformation. Normalizing provides stress relief and in addition increases the uniformity of the grain structure. Stress relieving of weldments is frequently performed by heating to about 650° C (1,200° F). While grain refinement is not obtained at this temperature, the chances for distortion are less than those that might be introduced by the allotropic transformation, which occurs at higher temperatures.

Welding Processes and Design 10

In the preceding chapter, the essential welding requirements of atomic closeness and atomic cleanliness were pointed out. It was noted, in the discussion of bond types, that while not always essential for welding, heat is an important part of most practical processes. Heat is necessary for fusion, heat makes metals become more plastic, and heat assists in obtaining cleanliness in many processes. The more important welding processes differ primarily in, and, in fact, are usually named for the heat source.

An integral part of practical welding processes is the method of obtaining and, of equal importance, of maintaining cleanliness in the weld area. Not only is it necessary to obtain atomic cleanliness for proper fusion but also the heated metal, particularly during fusion welding, must be protected from excessive contamination from the atmosphere.

HEAT FOR WELDING

Furnace Heat Not Localized. Energy sources used for welding are characterized by two important features: the degree of localization permissible and the rate of heat input possible. Heating in a furnace by radiation and conduction may permit a large total heat input but results in thorough heating of the entire part or assembly. This method would be unacceptable for fusion welding because melting of the entire weldment would occur but may be the preferred method for brazing and soldering. The base metal temperature being uniform, stresses caused by temperature changes are minimized. The ease of control makes furnace brazing adaptable to production quantities. Furnace heating is the usual method of preheating weldments to permit stress equalization and lessen the probability of cracking.

Forge Welding—a Pressure Bond. The process called *forge* welding is named for the initial method of heating in which the parts to be heated are placed directly in the fire of a forge, a special type of furnace. The parts, heated either locally or throughout, are then subjected to pressure (manual hammering in the case of blacksmithing) to produce the weld. The blacksmithing art is still important, but forge welding has been largely replaced by other methods.

Localized Heat Most Common. The most important welding processes make use of localized heating. For fusion welding, this is a necessity to prevent excessive melting and to restrict the heat-affected zone in the base metal. The temperature differential in the weld area will depend not only on the rate of heat input and the degree of localization but also on the thermal properties of the base metal and the geometry of the weldment. Heat sources differ in the maximum temperature possibilities, the degree of concentration, and in the maximum practical amount of energy that may be transferred.

Heat Source Influences Cleanliness. The choice of a heat source may be governed by the contaminating influence on the base metal. With some heat sources, especially those of chemical nature, the atmosphere to which the weld is subjected is determined by the heat source. With most types of electric heating, the atmosphere may be controlled exclusive of the heat source.

Economic considerations always play a large part in the final determination of a heat source. The actual energy costs, based on fuel or electricity, differ to some extent, but the choice is most frequently made on the basis of initial equipment cost, availability, portability, and the suitability of the process and equipment for the amount of welding and kind of material to be welded.

CHEMICAL REACTIONS

Oxyacetylene. The oldest and still most used source of heat based on a chemical reaction is the burning of acetylene (C_2H_2) and pure oxygen. A reducing or carburizing flame prevents or reduces decarburization and causes less oxidation of steel. An excess of oxygen produces a strongly oxiding flame that has only limited use but yields maximum temperatures. With three parts oxygen to one part acetylene, the temperature is 3,482°C (6,300°F). Other temperatures range from 815°C (1,500°F) at the tip of the inner cone of a neutral flame (one to one proportions of oxygen and acetylene) to about 3,300°C (5,972°F) in the hottest portions of the outer envelope.

Portability an Important Advantage. Oxyacetylene has advantages of portability, low first cost, and flexibility. With relatively simple equipment, operations ranging from brazing and soldering to flame cutting may be performed. For fixed installations and high production processes, the electric arc is used more than oxyacetylene because of the greater heat input that may be obtained and the lower cost of electrical energy.

Other Gases Less Used. Other gases burning with oxygen are also used but to a much more limited degree. Oxyhydrogen can provide a strongly reducing flame without the soot associated with oxyacetylene and is used for welding aluminum and lead. Natural gas, propane, or butane, burned with oxygen, are used for preheating and for brazing and soldering but have limited temperatures, making them less useful than oxyacetylene for fusion welding.

THE ELECTRIC ARC

Practically all production welding today makes use of electricity as an energy source. The first application was the electric arc, developed about 1880 (Figure 10-1) but restricted in use until the development of coated electrodes. The electric arc is one of

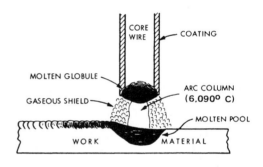

Figure 10-1
The welding arc

the hottest sources of energy available except for nuclear reactions. Arc column temperatures are near 6,090° C (11,000° F), which is well above the melting points of common metals and alloys. With typical arc-welding conditions of 25 volts at 300 amperes, the total energy suppplied would be 6,550 kilocalories per hour (26,000 BTU per hour).

Ionization Establishes Current Path. Most gases, including the atmosphere, are very poor conductors at room temperature, and the voltage necessary to maintain an arc over any practical distance would be very high. However, gas molecules at arc temperatures have such high velocities that they ionize (lose some electrons by collision) in numbers sufficient to make the gas highly conducting for electric current. When the arc is extinguished, it cools and loses its ionization in the order of one thousandth of a second, and reionization must occur before the arc can be reestablished.

The temperature of the arc is essentially constant throughout the length and diameter of the arc column. The electrical characteristics, including the voltage drop in the arc and at the surfaces at which the arc terminates, are determined by the composition and length of the arc. With long arcs and highly conductive gases such as hydrogen, higher inputs are required to maintain the arc.

The Work Frequently Serves as One Electrode. The arc usually exists between the work and a metal rod, which may progressively melt and serve as filler material or may be nonconsumable.

Some Metal Lost during Transfer. Welding arcs with consumable electrodes transfer this metal in molten form to the weld pool on the work. Transfer may be by fine metal spray or by relatively large globules and rivulets that may even short-circuit the arc temporarily. The rate of electrode burn-off is almost directly proportional to the welding current for any given rod diameter. However, the range of currents that may be used with any electrode to obtain a balance between burn-off and heating of the base metal is limited. From 10% to 30% of the melted rod is normally lost through vaporization and spattering outside the molten pool.

Gas Shielding Improves Quality. During transfer across the arc gap, the molten metal is shielded by protective gases from oxidation and other reactions with the arc atmosphere. These gases may be provided by the burning of coatings on the welding rod itself, by flux powders beneath which the arc burns, or by a flow of shielding gas from an external source.

Straight Polarity—Welding Rod Negative. With certain welding rods, the polarity of the rod with respect to the work exerts a measurable influence on burn-off rate and the amount of spattering. When the rod is negative, the setup is called *straight* polarity. When the rod is positive, the setup is called *reverse* polarity. Manufacturers designate the preferred polarities for most rods.

Arc Welding Versatile and Important. Arc welding has developed into the most versatile of all welding processes. Power supplies of almost unlimited capacity are available, and deposition rates in excess of 100 pounds per hour are used with the faster procedures. Many production processes have been developed, most with automatic regulation of current, rod feed, and speed of travel along the proper path. With proper shielding, most metals and alloys may be arc welded. Products that are regularly arc welded include tanks and other pressure vessels, structural steel, large diameter steel pipes, ship hulls and fittings, large machinery frames, and aircraft structures.

Percussive Welding. One other use of the electric arc is in *percussive* welding, a process more closely associated with pressure than fusion methods and used only for making butt joints between the flat ends of workpieces without filler material. The workpieces are connected to a large capacitor charged to about 3,000 volts, then driven toward each other by high spring or air pressure. Before contact can take place, arcs with current on the order of 50,000 to 100,000 amperes are established. These high currents quickly heat the surfaces of the work to vaporization temperatures. The vapor holds the workpieces apart until the capacitor is nearly discharged, at which time the pressure completes the contact against a thin film of clean molten metal. Equipment costs are high and applications are limited, but percussive welding may be used for joining widely dissimilar materials. Heat effects in the base material are limited in extent.

Stud Welding. A further variation in the use of an arc for welding is in the process called *stud* welding, developed in the shipbuilding industry for attaching steel studs to the steel deck of a ship. These studs are then used for holding the wood overdeck. The stud is supported in a special gun and forms the electrode in much the same manner as the filler material in conventional arc welding. It is then moved to the work until an arc is established, drawn back, then forced into the work—after a short period of arc heating—with sufficient pressure to cause some upsetting of the end of the stud. The process is used primarily for attaching threaded fastening devices in applications similar to that described above.

WELDING EQUIPMENT AND PROCEDURES

Most of the basic shape-producing methods make use of a relatively small number of equipment types for each of the individual processes. For both practical and economic reasons, the majority of welding processes make use of heat to establish the conditions necessary for welding. Most heating means are used at one time or another, so that the equipment design varies over a wide range. Welding is still in an earlier stage of development than casting, forging, pressworking, or machining, and new techniques with associated equipment are constantly being developed. At some future date, a higher degree of equipment standardization is likely, but at present, each new development adds another piece of specialized equipment.

ARC-WELDING ELECTRODES

Coatings Provide Protective Atmosphere. Early welding rods were bare iron wires, with which it was difficult to maintain stable welding arcs: the deposited metal was frequently porous or contained oxides and other inclusions. Modern welding rods for manual use are usually heavily coated with constituents that alleviate these problems.

The first function of the coating is to provide a gaseous shield that flushes away the atmospheric gases to prevent oxidation and other gaseous contamination of molten metal during transfer from the rod and after deposition in the molten pool. The gaseous shield generally also contains ionizing constituents that assist in ionizing the arc atmosphere by reducing the effective ionization potential so that the arc may burn with lower applied voltage. Sodium salts are commonly used for direct-current welding rods. Potassium salts are used for alternating-current welding rods for which arcs are more difficult to maintain because the current passes through zero 120 times each second (twice for each cycle of 60 hertz current).

Slag Protects Hot Metal. In addition, the coating may provide slag-blanket forming materials, which form a protective layer over the deposited weld metal. The insulating coating reduces the rate of cooling by heat loss to the atmosphere and protects the hot metal from atmospheric oxidation and gas absorption at the higher temperatures at which gases are readily soluble in the metal. For welding on vertical and overhead surfaces, special coatings with high slag viscosities are needed to prevent the slag from running off the surface of the metal during the period when the slag itself is molten.

Coating May Add Filler. In high-depositon-rate rods for flat position welding, extremely heavy coatings may be employed to carry powdered iron or iron oxide materials that combine with the deposited metal to add to the deposition rate. *Contact* electrodes are designed with coatings that burn off slowly enough to support the rod at a proper distance above the work for good arc length with less operator skill than demanded by the usual manual procedure. The operator merely drags the electrode over the work, yet maintains a good arc position as the coating burns away in unison with the melting of the metallic material.

MODIFICATION OF ARC WELDING FOR SPECIAL PURPOSES

Manual Procedures Very Versatile. Many installations in use today are for manual welding. Most use coated electrodes of consumable types; shielding of the arc is provided by burning of the electrode coating. The core wire provides the deposited metal. These electrodes are manufactured in stick form with core wires of various diameters and coatings for various welding purposes.

Manual Welding Economical For Small Quantities. Manual welding is costly in terms of time and labor as compared to automatic production processes but requires little or no setup time. Speed of manual welding is increased, where feasible, by using work positioners. These permit welding on complex shapes to be carried out in optimum welding positions, flat or horizontal if possible. In this way, high-deposition-rate electrodes may be employed to speed the work and lower its cost. Certain applications, such as repair and maintenance welding, construction of bridges and structures, and welding of cross-country pipelines, do not permit positioning of the work.

Even so, welding often proves to be far cheaper and produces more reliable structures than other fabrication processes.

Quality and Speed Improved with Modifications. When manual arc welding with stick rods cannot provide welds of high enough quality or when the nature of the work, especially the amount of welding to be done, permits higher setup and equipment costs with reduced operating labor time, a number of modifications are available.

Inert Gas Shield—Tungsten Electrode Welding of many modern metals and alloys, such as magnesium, titanium, stainless steels, and others is done with *gas tungsten-arc welding* (Figure 10-2). In this process, first developed during World War II for welding magnesium alloys, an arc is maintained between a nonconsumable tungsten electrode and the workpiece, while shielding is provided by an inert gas or gas mixture, most commonly argon or helium. Filler metal may or may not be added as the particular application requires. This method has been well developed and finds many applications today, particularly for welding some of the difficult materials. In the past, this nonconsumable process has been referred to as tungsten inert gas welding.

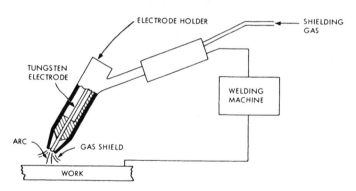

Figure 10-2
Schematic diagram of gas tungsten-arc welding

Wire Electrode May Supply Filler. Several variations of *gas metal-arc welding* (Figure 10-3) have been developed. Processes of this type have in common the use of a filler material in wire form, which is continuously fed into the weld metal pool, and a shielding gas, or mixture of gases, to provide the protective atmosphere. Filler-wire diameter may range from 0.5 to 30 millimeters (0.020 to 0.125 inch), and currents may range from 90 to 800 amperes. Equipment is available for both hand-held and machine-guided operation.

Several Gases Used as Shields. Argon, helium or mixtures of argon and helium are the commonest shielding gases, particularly for high alloy steels and nonferrous metals, because of their complete

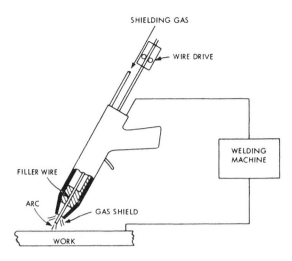

Figure 10-3
Schematic diagram of gas metal-arc welding

chemical inertness. However, the gas mixture has considerable effect on the depth of penetration, the contour of the weld surface, and the arc voltage. From 0.5% to 5% oxygen is sometimes added to improve the weld contour. Carbon dioxide gas is frequently used when welding mild steel; even then it is difficult to avoid porosity in the weld. Weld quality may be improved by providing a small amount of dry flux as a magnetic powder that either clings to the rod as it emerges from the holder or is contained in the center of hollow filler-wire. Similar improvement may be obtained by using two shielding gases: a small amount of inert gas such as argon or helium near the rod and a larger flow of cheaper carbon dioxide surrounding the inert gas.

AUTOMATIC WELDING

Almost all electric arc processes except those using covered electrode wire are amenable to a certain amount of automatic or machine control. Those using gas metal-arc and/or flux core wire with gas shield are sometimes adapted for automatic operation. Usually a constant voltage power supply and adjustment of current flow permits the burn-off rate to maintain an approximately uniform arc length regardless of the rate of wire feed.

The automatic feature of this kind of welding is often the result of experimentation and custom design to provide proper coordination in the relative motion between the heat source and weldment. Sometimes, the electrode holder is moved through a predetermined path by control of a holding fixture that is clamped to the work. In other cases the arc may remain in a fixed location and the weldment moved past it by action of the positioner on which it is mounted. Although in a root pass a straight line motion is most likely to be used, on multiple pass welds, a waving motion such as might be imparted by an expert welder may be incorporated by automatic control to improve the weld quality and increase the deposition rate.

Gas Tungsten-Arc Welding. A great amount of automatic welding is performed by the gas tungsten-arc method becuase the non-consumable tungsten electrode provides for a heat source with good stability. Most automatic gas tungsten-art welding machines are applied with electronic controls that automatically move the electrode holder upward or downward to maintain a constant arc length. The possible compactness of the electrode holder permits use of the method in locations where a human operator could not see or manipulate. Much of the development work was performed in submarines for successfully welding pipe in inaccessible areas.

The method is widely used for pipe welding both in the field and in the shop. In some cases, welds without filler wire are produced—most on relatively thin sheet metal products. In other cases, cold filler wire is fed into the weld puddle or in still other cases, preheated filler wire is fed to promote faster welding. Recently developed are pulsed-arc power supplies capable of providing various pulse characteristics to the arc. Such controls impart high frequency agitation of the matter puddle, in effect stirring the oxides and evolved gasses out of the weld.

As with any automatic process where the human element is reduced, reliability and consistency tend to be improved once the process is in operation. However, because the operator no longer has the ability to instantly compensate for observable errors, problems of initial preparation and fit-up become more critical.

Automatic Welding Under Flux. A high production process in wide use today is *submerged arc welding* (Figure 10-4). The power supply and feeding arrangement are similar to those that would be used with gas metal-arc welding, but shielding is provided by a granular flux fed from a hopper to surround the arc completely. Part of the flux is fused by the heat of the arc to provide a glassy slag blanket that protects the molten metal and the solidified weld as it cools. In addition, the normally nonconductive flux becomes conductive when fused and permits very high current

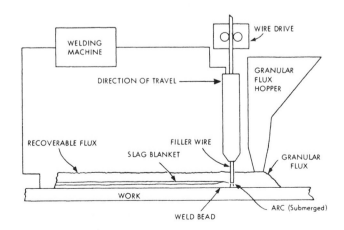

Figure 10-4
Schematic diagram of submerged arc welding

densities that give deep penetration. Because of the greater penetration with a saving of filler material and a higher welding speed for a given current, smaller grooves may be used for joint preparation with this process than with others. It is basically a shop process.

ELECTRIC RESISTANCE HEATING

Heat for an important group of hot pressure-welding processes is supplied by the passage of electric current through the work. The rate of power expenditure in any electrical circuit is given by

$$P = I^2 R$$

where P is the power in watts; I, the current in amperes; and R, the resistance in ohms.

Highest Resistance at Interfaces. Heat is generated throughout the circuit, and resistance-welding processes are based on the fact that the highest resistance occurs at the interfaces between metal surfaces where the contact is limited to a number of points of relatively small area. This condition occurs not only at the interface between the workpieces, where maximum heat is desired, but also at the contacts with the electrodes, for which the heating effect is minimized by using high conductivity copper alloys with water cooling and high pressure contact of formed surfaces.

Melting Incidental Only. As the contact points heat between the work surfaces, they become plastic, and the clean metal union is expanded by deformation and by the fragmentation, spheroidization, and diffusion of the oxides into the base metal. Some local melting may take place but is not necessary for the process to be successful. Even with the increased area of contact, the interface area remains the point of greatest heat generation because the resistance of the base metal rises as its temperature is increased. The duration of the current is controlled by a timer that in most cases regulates the periods of current flow by controlling the number of cycles of alternating current permitted to flow through the primary of the step-down transformer. The pressure is also timed, with an increase to cause plastic flow after heating has occurred.

Dissimilar Metals May Be Joined. Nearly all metals, as well as most combinations of different metals, may be resistance welded. Difficulties are sometimes encountered in welding high conductivity metals such as aluminum and copper or in joining parts of different thicknesses. Experimentation to establish the best weld conditions will produce satisfactory welds for most applications.

Spot Welding for Joining Sheet Metal. The most important applications of resistance heating are for *spot welding* and its variations. Used primarily for lap joints between flat sheets, spot welds are obtained by concentrating the pressure and current flow with

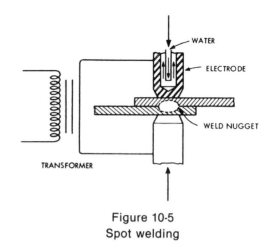

Figure 10-5
Spot welding

shaped electrodes, as shown in Figure 10-5. Accurate control is necessary to prevent burning of the electrodes and excessive heating of the base material, which would cause too much plastic flow under the pressure of the electrodes. Spot welding is sometimes facilitated by interrupting the current flow and using a series of short heating periods to provide a different heat distribution.

Modified Spot Welding. The two most common variations are *seam* and *projection* welding, shown in Figure 10-6. In seam welding, a series of overlapping spot welds produce a continuous joint used primarily where pressure or liquid tightness is a requirement, as in automotive gasoline tanks.

In many cases, multiple spot welds or single spot welds of highly localized character may be made by confining the area of contact to projections on the surface of one or both workpieces. Large electrodes shaped to the contour of the work may be used, and

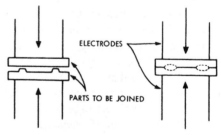

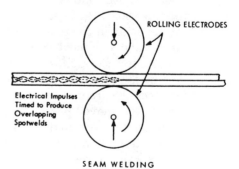

Figure 10-6
Variations of spot welding

the exterior of the part has little or no marking from the electrodes. Uses of *projection* welding include the joining of electrical contacts to relay and switch parts and the manufacturing of fencing in which the projections are inherent in the product where the wires cross.

Spot Welding an Important Assembly Process. Spot welding and its variations are among the most used joining processes in the manufacture of high quantity goods, such as automobiles, home appliances, office equipment, and kitchenware. Dissimilar metals and parts of different thicknesses may be joined. Little cleaning of the parts is necessary either before or after welding. The greatest limitations are the initial cost of equipment, the experimentation sometimes necessary with new applications, and the restrictions to joining relatively thin material except in the case of projection welding.

SPECIAL WELDING PROCESSES

As in the case of sheet-metal forming, a number of limited-use joining processes have been developed for special applications. These may be concerned with the welding of refractory or easily oxidized metals, of metals that require extremely high rates of heat input, or of heavy sections or may simply involve special procedures that assist some otherwise conventional process. Most are of rather limited use because of the special equipment required, the restriction of sizes, the high cost involved, or being new, the lack of widespread knowledge.

ELECTRON-BEAM WELDING

Energy for heating may be made available in many forms. In the *electron-beam gun* (Figure 10-7), a stream of high energy electrons is focused electrically toward a spot on the surface to be heated. Rapid localized heating takes place with the possibility of melting for welding or of complete vaporization for removing metal. The process is carried out in a vacuum so that no products of combustion and no contamination or oxidation of the heated work occur. The boiling of the molten metal at the high temperatures removes impurities that may be present, and the resulting weld may be of higher quality than the base metal. The high rate of heating restricts the heat-affected zone, and there is minimum distortion and alteration of physical properties. A ratio of fusion depth to width of as much as twenty is possible.

The process uses high cost equipment, and the total amount of heat available is small. Electron-beam welding is valuable for welding beryllium, molybdenum, zirconium, hafnium, and other refractory metals difficult to weld by other methods.

PLASMA ARC

For most gases, the stable molecular form at room temperature contains two atoms, but the gas, when ionized, becomes *monatomic* in form. A *plasma* is a gas that has been heated to such a temperature that the gas is ionized. A reduction in temperature results in the recombination of atoms to the molecular form and the release of energy as heat. The gas column in arc welding is ionized, but in this case, it is a relatively small, stationary quantity of gas that is involved.

In the *plasma-arc* process, a stream of gas is ionized by heat as it is passed through an electric arc by one of the two methods shown in Figure 10-8. Thermal expansion of the gas stream causes it to flow at supersonic speeds as its diameter is restricted by the magnetic properties of the arc. The drop in temperatures caused by contact with the relatively cool work surface results in loss of ionization and the release of large amounts of heat directly at the surface to be heated. The process has a high intensity and a high rate of heat transfer, which makes it useful for welding high conductivity metals such as aluminum.

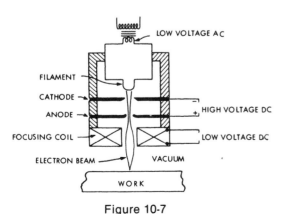

Figure 10-7
Schematic diagram of a simple electron-beam gun

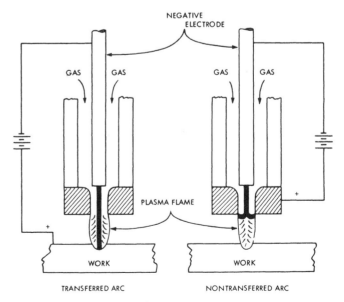

Figure 10-8
Plasma arcs

ULTRASONIC WELDING

Vibration Aids Cleaning. One of the principal limitations on cold bonds is the excessive deformation required to provide enough fragmentation of the oxide layers on the contacting surfaces. Cold bonding may be performed with less deformation by applying high frequency mechanical energy in the process called *ultrasonic* welding. The vibrations introduce shearing forces that assist in the fragmentation; as a result, more than 50% clean-metal contact may be established.

Both spot and seam welds may be made, and the widest use has been for metals difficult to join by conventional processes. These include stainless steel, molybdenum, zirconium, various bimetal combinations, and thin foil or sheet aluminum. The upper limit is about 0.100 inch, although thin sheets may be welded to thicker sections. Ultrasonic welding is also an important assembly method for plastics.

FRICTION WELDING

In ultrasonic welding, mechanical energy is supplied to facilitate fragmentation. In *friction* welding (Figure 10-9), mechanical energy is supplied not only to facilitate fragmentation but also to develop heat. In the process, used almost exclusively for making butt welds in heavy round sections, the bars or tubes are brought together with high force while one is rotated. The friction develops sufficient heat to make the metal plastic and to permit cleaning and closeness to be achieved in much the same manner as in resistance butt welding.

ELECTROSLAG WELDING

Slag Protects the Heated Metal. The principal of *electroslag* welding is illustrated in Figure 10-10. The edges to be joined are placed in a vertical position with a gap between them. Water-cooled copper shoes or slides cover the gap where the welding is in process. Slag is first deposited in the gap and a wire electrode introduced to form an arc. Once the arc has melted the slag, the arc is automatically extinguished, and heat is produced by the passage of current through the molten slag. The electrode is fed into the slag as it melts, and, as the gap fills, the copper shoes and electrode guide are gradually raised. The process might well be defined as continuous casting, with the base metal and the copper slides forming a moving mold. The slag forms a protective layer for the weld pool and, in addition, forms a coating over the copper slides that protects them from the molten metal. By changing the rate of wire feed and the electrical input, the rate of deposition and the penetration into the base metal may be controlled.

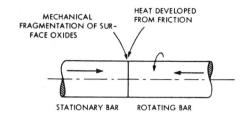

Figure 10-9
Friction welding

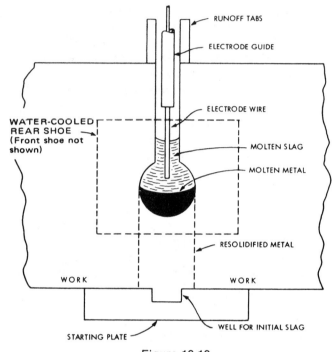

Figure 10-10
Electroslag welding

Multiple Electrodes Needed for Heavy Sections. A single electrode is used for sections up to about 2 inches thick. For thicker sections, multiple electrodes may be used, and melting rates of up to 40 pounds per hour for each electrode are possible. While the principal applications have been for forming butt welds between plates and for producing heavy-walled cylinders rolled from flat plate, shaped rather than flat slides may be used for producing tee joints or special built-up shapes on the surface of a part. In a newer variation of the process, an arc is used continuously, without slag, but with a protective gas atmosphere fed through ports at the tops of the copper slides.

EXPLOSION WELDING

In recent years, explosion welding has developed into an important process particularly suited to joining large areas of two or more metals of different compositions. Standard explosive materials supply the energy to produce the weld, which may be made on flat or curved surfaces.

Progressive Cleaning and Welding. A uniform covering layer of explosive material is detonated to produce a shock wave that progresses uniformly across the material to be welded. The materials to be welded are originally spaced a small distance apart. The shock wave from the explosion closes the gap in such a way that surface impurities are pushed ahead and extremely high pressures establish the contact of clean metal for welding.

The greatest use for the procedure is in coating, or cladding, structural metal with a more expensive but more corrosion-resistant metal. The purpose may be to protect the metal from ordinary environmental exposure or to prevent damage from more intense exposure such as in chemical process containers.

DIFFUSION WELDING

Solid-state processes of joining metals were the earliest used and antedate the fusion processes. Revived interest in the principles of the solid-state processes, however, has recurred in very recent years together with increased theoretical knowledge of solid-state bonding. The result is development of diffusion welding.

Pressure, Temperature, Time—Independent Variables. The process involves the establishment of a smooth, clean surface that must be maintained until the weld is accomplished. This often means protecting the surfaces in an inert gas environment for a few seconds to a number of minutes. Low to moderate pressure is applied to the surfaces to be joined at the same time the temperature is raised. The welding temperature is somewhat dependent on other conditions but usually falls someplace between the recrystallization temperature and the melting temperature of the material.

Present Use Limited. Diffusion welding does not at present seem to be economically competitive with other processes when the other processes can produce satisfactory results. The main use to date has been in welding new materials to avoid metallurgical, corrosion, and physical problems sometimes associated with older welding techniques. The process has been used most for joining special alloys in aerospace and atomic energy applications.

WELDING DESIGN

Welding may fulfill either one of two basically different design concepts. As a basic shape-producing means, welding competes with other basic processes, especially forging and casting. The individual parts making up a weldment are most frequently cut from rolled sections that are produced in high quantities at low cost. Ideally, the finished weldment may be thought of as an homogeneous structure equivalent to a single part. Even with less than 100% joint efficiency, the single-piece concept may still be used in design if the maximum permissible stresses are considered and joint areas are increased where necessary.

Unitized Product. The single-piece concept is used in many applications. For example, in much welded pipe, the weld is undetectable without critical examination; a drill or reamer shank is continuous with the body of the tool, even though they are of different materials. In modern welded structural steel assemblies, the joints may be stressed as continuations of the beams involved, although strengthening plates are sometimes necessary. In many instances, welding permits the single-piece concept to be applied to designs that would not otherwise be possible.

Assembly Fastening. The second concept of welding is as an assembly means in competition with mechanical fastenings. The welded assembly is generally permanent, but the individual parts retain their identity, and the strength of the structure is frequently governed by the strength of the joints. The use of spot, seam, and projection welding is normally in this class. In many cases, not only are the mechanical fasteners eliminated but also preparation by drilling or punching holes is unnecessary and gaskets are no longer needed for sealing. Fitting of parts together may be simplified because alignment of holes is not required. The parts may be merely positioned with proper relationship to each other.

JOINTS

The terminology applied to the shapes of welded joints is somewhat loose. The type of joint and the type of weld are two different considerations. Two flat plates, for example, may have their edges butted together, one may be lapped over the other, or they may be placed at right angles to each other. The configuration adopted would be referred to as the type of *joint*. Although some joints are more conviently welded by some processes than by others, and some processes are restricted to certain types of joints, the specification of a joint type does not automatically specify the welding process or the manner in which filler material is to be placed.

Weld Type Usually Distinguished from Joint Type. The actual shape of the bonded area or the cross-sectional shape of the filler material, frequently governed by the preparation given the edges of the part to be welded, is known as the type of *weld*. In the lapped position, the plates might be joined by building up fillets along the edges, filling in holes or slots in one plate with weld metal, spot welding, or seam welding. Frequently, a close connection exists between the type of weld and the process that may be used. Either term, *joint* or *weld*, is sometimes used to refer to both the relative positions of the parts to be welded and the type of weld.

Figure 10-11 shows the weld types that may be produced by fusion welding. Following the name is the drawing symbol for each type. Bead welds are often used for building up metal on a surface where joining is not needed. The type of groove weld used will have considerable influence on the penetration into the base metal necessary for good bonding and on

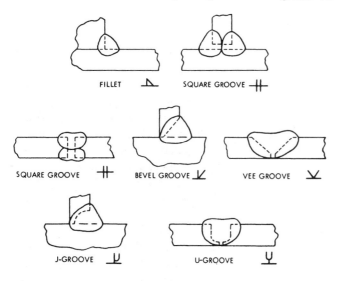

Figure 10-11
Fusion-weld types

the amount of distortion encountered. A vee or bevel weld requires simpler preparation than a U- or J-weld but results in greater distortion because much more heat is present at the opening of the vee than at the bottom. The heat difference is not so great in a U-weld. Where access is available to both sides of the members, many groove welds are made in double form, especially for heavy members. Adequate penetration with square grooves is generally possible only by welding from both sides. The weld types shown also apply to braze welding, except that in this application no melting of the base metal would occur, and the dotted lines in the figure would be the extent of fusion.

Configuration Determines Joint Type. Five basic types of joints are used for welding. These are shown in Figures 10-12 through 10-16. The types of welds that may be used with each and the standard weld symbols that apply are shown.

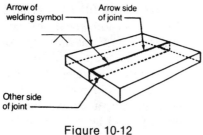

Figure 10-12
Butt joints

The Welding Symbol. Figure 10-17 illustrates the elements of a welding symbol as recognized by the American Welding Society. The symbol is used on

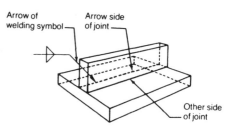

Figure 10-13
Tee joints

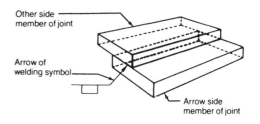

Figure 10-14
Lap joints

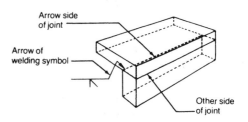

Figure 10-15
Corner joints

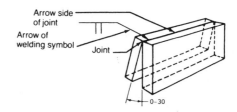

Figure 10-16
Edge joints

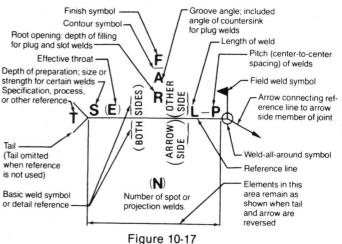

Figure 10-17
Elements of the welding symbol

drawings to designate the details of a weld. Any part of the symbol that is not needed for clarity may be omitted. Figure 10-18 shows the manner in which the symbol would be used to describe a welded corner joint together with the result of following these specifications. The joint is to have a 1/4-inch unfinished fillet weld on the inside of the corner (opposite side) and a 1/4-inch back weld on the outside (near, or arrow side) that is to be ground flat. The shielded-metal arc-welding process is to be used. It is to be a continuous weld along the corner because no pitch or spacing is designated.

DESIGN CONSIDERATIONS

It has been fully realized in recent years that welding is a unique process and that all of the design rules applied to other processes do not necessarily apply to

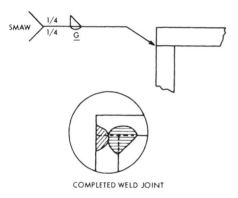

Figure 10-18
Example of welding symbol use

welding. Welding started as a repair method and developed from this, primarily as a substitute for other methods of *joining*. When it is used strictly as a joining method, particularly by spot welding, little trouble is experienced. However, when parts are fully joined to form rigid, *one-piece* structures, designers have not always realized that such structures do not respond to loading in the same way as a bolted or riveted structure. Many structures must allow for yielding or shifting in service that might be permitted by a bolted structure but not by a weldment, unless the design were changed.

Unit Structure—Special Consideration. A number of failures of welded ships and storage tanks have been traced to cracks that can grow to a large size in a welded structure but would be interrupted by a mechanically fastened joint. Monolithic welded structures have been found to be somewhat more notch sensitive with a corresponding drop in impact strength, especially at low temperatures.

On the other hand, designers have not always taken full advantage of the potential joint strengths offered by welding. Welding can produce rigid joints that improve beam strengths. The material would be used inefficiently if a welded structural steel assembly were designed according to rules that permit freedom in the joints as is generally assumed for bolting or riveting. Large improvements in joint strength and ductility have resulted from improved methods for preventing contamination of the weld metal and as the metallurgical changes that take place in a weld have become fully understood.

WELDABILITY

Weldability Varies with Material. The relative ease with which a sound union may be produced between two parts by welding is knows as the *weldability* of a metal. A number of factors must be considered. Some metals may be more easily contaminated than others. The contamination may consist of gross oxide inclusions or voids that would be very apparent in a cross section of the weld, or of microcontamination that results in structural changes detectable only by examining the metallurgical structure. Gross defects not only reduce the actual cross section of the weld but also introduce stress concentrations that are particularly harmful in a metal with low ductility. The principal effect of structural changes is reduced ductility. Contamination can be controlled by providing the correct environment for the molten metal.

Hardenability. Especially important for steels is consideration of the hardenability of the metal. It will be remembered that this term is related to the cooling rate necessary to form a structure of given hardness in a steel. Again remembering that as hardness is increased, ductility decreases, the effect of hardenability on weldability can be predicted. In all the important welding processes, the metal is heated near or above the melting temperature, and cracking or high residual stresses as the metal cools differentially can be prevented only by yielding of the metal in the weld area. With few exceptions, any element that is added to pure iron increases its hardenability and therefore decreases its weldability by reducing ductility and increasing the possibility of cracks or high residual stresses. Therefore, increased welding difficulty can be expected as carbon or alloy content is increased in any steel. The major exception to this rule is the addition of vanadium, which reduces hardenability.

Thermal Conductivity. Another factor affecting weldability is the thermal conductivity of the metal. If a metal had infinitely high thermal conductivity, it could not be fusion welded at all because it could not be locally melted. Aluminum, for example, has such high conductivity that high rates of heat input are required to prevent excessive melting of the base metal. On the other hand, stainless steels have low conductivity, which results in hot spots, and very high temperature gradients in the weld zone, which results in increase of the stresses developed on cooling.

Composition. Composition can have other effects than those on hardenability. Stainless steels may not be hardenable to martensite at all but may develop higher stress on cooling than carbon steels of equivalent strength at room temperature because stainless steels have higher yield strengths at elevated temperatures. The chromium in stainless steel is especially subject to oxidation, and chromium oxide does not separate out easily from the molten weld pool. Many nonferrous alloy constituents are subject to segregation when cooled rapidly.

Recrystallization. Heat produces other effects on structure than those of quench hardening of steels. Material that has been cold worked is automatically recrystallized during welding, usually for a considerable distance away from the actual weld. Most aluminum alloys begin to recrystallize at about 150° C (300° F) so that a weldment made from work-hardened aluminum may actually be more ductile in the heat-affected zone than in the unheated base metal but only with an accompanying reduction of strength. Grain growth will follow recrystallization, and even subsequent heat treatment cannot restore a desirably small grain size in most nonferrous metals.

Corrosion Resistance. The corrosion resistance of stainless steels may be especially affected by welding. At low cooling rates, small amounts of carbon can combine with chromium and reduce the corrosion resistance. Nearly all cooling rates will exist somewhere in the weld area; consequently, corrosion resistance will likely be lowered in some spots. Post-heat treating of stainless steel weldments is nearly always required to restore maximum corrosion resistance.

In addition to the structure effects, heat causes other changes. The surface of practically all metals is oxidized at welding temperatures. While surface oxidation may not directly affect strength, it does affect appearance and may produce surface imperfections that lead to fatigue failures or serve as focal points for intergranular corrosion.

Distortion. Even when the residual stresses do not lead to actual failures, they cause other difficulties. The dimensions of a weldment are usually different before and after welding, and machining is nearly always necessary for close dimensional control. The machining itself may release residual stresses to cause further dimensional change. When close tolerances must be held, stress relief prior to machining is usually required.

A number of precautions and corrections can alleviate the problems caused by stresses and distortions. If the amount of distortion can be predicted, the parts to be welded may be purposely off-positioned before welding to compensate. This procedure is somewhat like overbending sheet metal to compensate for springback. Some automatic compensation will occur in a double-groove weld made from both sides of a joint, but the first side welded usually will have the greatest effect. When a number of welds are to be made at a number of locations in a weldment, distortion may be controlled by choosing the proper sequence for making the welds.

Pre-Heat and Post-Heat Treatment. The most universal solutions to the problems of stresses and distortion are pre- and post-heat treatment of weldments. Pre-heat treatment does not eliminate shrinkage and yielding that lead to stresses, but by lowering the yield strength of the base metal, it provides a greater volume through which the shrinkage may be distributed, and by lowering the thermal gradients in the weld zone, it reduces the size of the stresses by distributing them over greater areas. Post-heat treatment relieves stresses by permitting yielding to occur at reduced stress levels; it can also help restore a uniform structure with an improved grain size, particularly in steel.

When materials have sufficient ductility, correct dimensions can be established by straightening. This may involve pressing operations in fixtures or localized heating with torches.

The factors that lead to residual stresses and distortion generally have an adverse effect on the strength of welded metals. Inclusions or voids not only reduce area but also are stress concentration points. Composition changes in the weld area may either increase or decrease strength with a corresponding change in ductility. In some nonferrous alloys, brittle intermetallic compounds may form that have a serious effect on ductility.

Weld Penetration. The efficiency of a fusion-welded joint may depend on the amount of *penetration* achieved. Although melting of the base metal is not absolutely necessary for bonding, and, in any case, proper bonding requires only that the surface of the base metal be melted, practical joint shapes cannot generally be heated to melting only on the surface. To obtain proper bonding at the bottom of a square-groove weld with most heat sources, it is necessary to melt a considerable amount of base metal. Heat sources differ in their ability to penetrate, that is, in the depth-to-width ratio of the molten zone that may be produced, dependent largely on the degree of heat concentration.

WELD DEFECTS

Many of the possible weld defects have been discussed, or indicated, earlier in this and the previous chapter. The following discussion is for the purpose of summarizing those most important and most likely to require the use of NDT.

The general sources of weld defects include: improper design, poor joint preparation, defects in the parent material, improper welding technique, faulty solidification of molten metal, and heating or cooling effects on both the base metal and the weld metal. Some depreciating faults, such as decreased strength of cold rolled steel due to recrystallization of the base metal in the heat affected zone, are inherent in the process and essentially become design problems. If the somewhat broad assumption is made that the design is proper, many defects are the result of improper welding technique. It follows then that an experienced, knowledgeable operator using care and good equipment should turn out the work containing the fewest defects. Even under the best of conditions, however, perfect results should never be expected. There are too many possible reasons for defects to occur. All *critical* welds require nondestructive testing for assurance of quality or as a means to enable repairs to be made.

FUSION WELDING

When welding is used during the manufacture of consumer products and for large structures, with the exception of resistance spot welding, a fusion arc welding process is most likely selected. The American Welding Society categorizes weldment defects in three general classes:

1. those associated with drawing or dimensional requirements
2. those associated with structural discontinuities in the weld itself
3. those associated with properties of weld metal or welded joint.

DIMENSIONAL EFFECTS

Warping. Differential heating and cooling sets up unequal stresses in the weld area that must be absorbed by position shift (warping), deformation (plastic flow), or cracking if neither of the others can occur. Although warping is inherent in the process, it can be minimized by proper welding control including joint preparation. When necessary, fixtures may be used also to minimize distortion. In some cases, peening to produce localized deformation or post-heating to equalize residual stresses may be needed to prevent cracking.

Weld Dimensions and Profile. Usually the unit strength of weld material is weaker than the unit strength of the base material it joins. This is due not only to the chemical composition normally used but also to the possible defects it may contain as a finished weld filler. When full strength is desired welds are made slightly oversize with a given shape. If the convexity is too large though, time and material are wasted and the chance of other defects is increased.

As can be seen in Figures 10-21, 10-22, and 10-23, some profile defects create discontinuities in addition to not doing their full job of reinforcement.

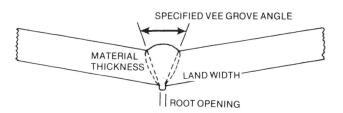

Figure 10-19
Warpage, angular distortion of a butt joint produced with a single vee preparation

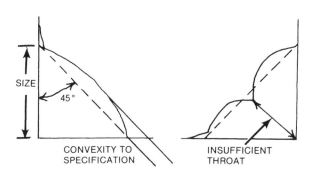

Figure 10-20
Fillet welds.
Ideal *top left* with others showing typical defects

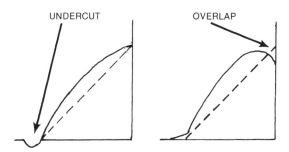

Figure 10-21
Double vee welds.
Ideal *top* with others showing typical defects

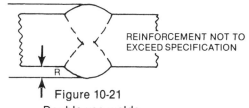

Figure 10-22

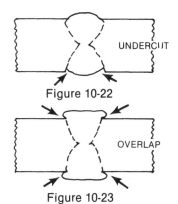

Figure 10-23

Final Weldment Dimensions. All weldments are designed to meet dimensions necessary to function properly or unite with other parts. Welds, especially when multiple, must be carefully controlled regarding spacing for overall dimensions to be within usable range. Accumulation of weld size error affects overall dimensions and even when balancing may cause poor quality welds.

STRUCTURAL DISCONTINUITIES

Porosity. The term *porosity* is used to describe pockets or voids that are the result of the same kind of chemical reactions that cause similar defects in castings. Gases are produced or released at high temperatures and when unable to escape, remain in the solidified metal. They may be microscopic in size or exist as large as 1/8-inch or more in diameter. It is seldom that porosity in welds can be eliminated completely but a few small, scattered pores may not create significant harm except in the most critical applications. As shown in Figure 10-24, porosity may exist as uniformly scattered, clustered, or linear.

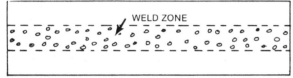

UNIFORMLY SCATTERED
Tend to be uniform size for a given condition

CLUSTERED
Often associated with some welding condition change

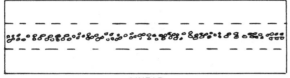

LINEAR
Occurs most often in root pass of a multipass weld

Figure 10-24
Three types of weld porosity

Inclusions. The most common inclusions that appear in welds are slag, metal oxides, and non-metallic solids that are entrapped during welding. They are to some degree associated with certain types of welding but are most likely to be present when the weld metal temperature has not been high enough to permit their floating to the surface, when there is an undercut or recess over which welding is performed, or almost certainly when insufficient clearing has been performed on previous passes of multipass welds. In welds made with an inert covering gas, inadequate fast flow or excess moisture in the gas can result in oxide and porosity formation. Figure 10-25 shows possible locations of slag in a multiple pass vee weld. In welds made by the fast tungsten-arc process, small bits of tungsten are occasionally dislodged from the electrode and enter the weld metal.

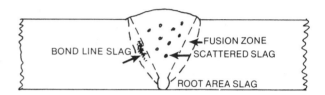

Figure 10-25
Some types and locations of slag inclusions

Incomplete Fusion and Inadequate Joint Preparation. Incomplete fusion can occur in any location where the base metal, or previous pass weld metal, has not been brought up to fusion temperature. Inadequate joint penetration, when present, usually occurs in the root area of the weld and is caused by similar reasons—sufficient heat for fusion does not reach the bottom of the groove. Either may be caused by welding operator error but inadequate penetration may also be caused by too close fit up or other improper joint preparation or design. Other contributing factors are too large electrode, too fast travel, or too low welding current. Figure 10-26 and 10-27 show some examples of poor fusion and penetration.

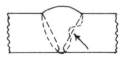

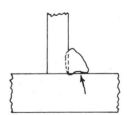

Figure 10-26
Incomplete fusion

Undercut. Undercuts are the result of melting base metal and not replacing it with weld metal, leaving a notch or groove. When occurring on the last pass of a multiple pass weld, or with single pass welds, the groove if deep may be a serious defect that should not

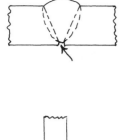

Figure 10-27
Inadequate penetration

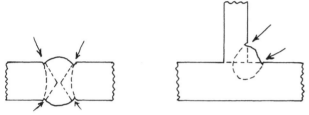

Figure 10-28
Undercuts in a double vee butt weld and in a horizontal fillet weld

be left. Undercuts such as shown in Figure 10-28 are results of operator technique in most cases but are also influenced by welding conditions.

Cracks. The cracking of weld metal and base metal in or near the weld zone is usually caused by high stresses set up by localized dimensional changes. Such changes are caused by the large thermal gradients established during heating and cooling of a weld joint. Cracking may occur during welding, during cooling, or particularly with hard or brittle materials at some later time. Weld cracks are most likely to occur when weldments are of heavy sections creating a faster quenching action.

Cracks in the weld metal are primarily of three types—transverse, longitudinal, and multiple star-shaped crater cracks, all of which are pictured in Figure 10-29. Sometimes the cracks are highly visible, sometimes magnification is required to see them, and at other times they can be detected only by nondestructive testing methods.

Crater cracks may be single or star-shaped multiple and form during shrinkage of the final weld pool. They mat propagate into longitudinal cracks or they may appear any place along a weld where welding has been stopped and restarted unless they are completely remelted in the process. Unless removed or fully remelted, cracks in the root of a weld are likely to propagate through all subsequent weld layers.

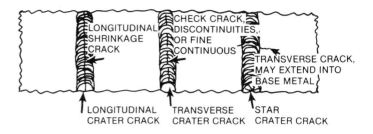

Figure 10-29
Types of weld metal cracks

Cracks in the heat-affected zone of base metal occur almost entirely only in metals that are heat-treat hardenable. Most are longitudinal in direction and sometimes may be extensions of bond-line cracks as shown in Figure 10-30. Crack defects in fillet welds may appear in the weld or as longitudinal toe cracks or longitudinal root cracks as illustrated in Figure 10-31.

Surface Irregularities. Occasionally surface irregularities and imperfections may be nuclei of future failure but usually they have little significance in weld joint strength and utility. However, spatter, weld ripple, uniformity of bead, and other surface qualities are frequently covered by specification and may require inspection even if their only effect is on appearance.

WELD METAL AND BASE METAL PROPERTIES

Weld Metal. The properties of the weld metal are controlled basically by the weld filler material and the way it is deposited. Most tests to determine its quality are destructive types and can be used only as spotchecks. Any weld inspector should however observe and check that proper materials and methods are used and that the welding operator uses the techniques necessary to produce desired qaulity.

Figure 10-30
Bond line crack extending into base metal

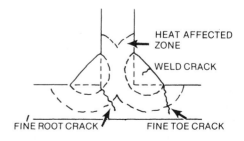

Figure 10-31
Root and toe type cracks in base metal with fillet welds

Base Metal. Similar to weld metal tests, most tests for checking properties of base metal are destructive. Code colors and other methods are used to identify that proper materials are being used. The inspector should also be fully aware at all times while performing nondestructive tests for weld quality that defects in base material may be indicated. Defects such as depicted in Figure 10-32 may have been missed in base material that was previously not inspected. Welds deposited over already existing defects can cause the base metal defects to enlarge or extend into the weld deposit. Such conditions found during weld inspections often indicate the need for more complete inspection of the base material prior to welding on subsequent weldments.

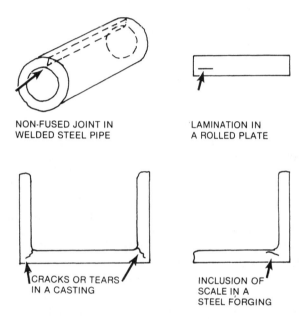

Figure 10-32
Some typical base material defects

BASIC SYMBOLS FOR NDT

In the interest of saving space and simplifying drawings and specifications, abbreviated symbols are accepted and encouraged to describe standard nondestructive test procedures. The American Society for Nondestructive Testing recognizes the following symbols:

Type of Test	Symbol
Acoustic Emission	AET
Eddy Current	ET
Leak	LT
Magnetic Particle	MT
Neutron Radiographic	NRT
Penetrant	PT
Radiographic	RT
Ultrasonic	UT

These symbols are used on a drawing with a testing symbol very much like the welding symbol used to specify welding types and procedures. As with the welding symbol, the placement of the basic test symbol below the reference line means testing is to be performed from the side to which the arrow points. Figure 10-33 shows the testing symbol which carries the tail only if some special reference is to be indicated and may at times be combined with the welding symbol for the same joint by carrying two reference lines.

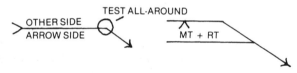

Figure 10-33
The testing symbol. (a) General form. (b) Combined with welding symbol to indicate that a vee groove butt weld is to be both magnetic particle and radiographically inspected from the opposite side.

Plastic Flow 11

Extensive use is made of the ability of most metals to undergo considerable amounts of plastic flow. The importance of the manufacturing processes based on plastic flow may be realized by considering some of the products. Of the total annual United States production of 100 million tons of steel, about 10% is used as castings, and the other 90% undergoes deformation of some sort, starting in nearly every case with a hot-rolling operation. For most products, additional hot-rolling or forging operations will involve plastic flow. More than 25 million tons are produced as cold-rolled plate, or sheet, which becomes the raw material for pressworking operations in which additional plastic flow produces most of the high production consumer goods, such as automobiles and home appliances. These 25 million tons are more than the total of all nonferrous metals and plastics produced annually.

Deformation Offers Unique Advantages. Numerous factors account for the use of deformation processes. When the quantity is sufficiently high to justify the extensive and costly tooling, many shapes can be more economically produced by deformation processes. One outstanding reason for this is the difficulty of casting very thin sections. Perhaps even more important is the high duplication accuracy of most deformation methods, particularly those in the cold-working category.

Important Properties Improved. The properties of wrought materials are in general much improved over their cast counterparts. Rolling, forging, and drawing generally tend to improve both strength and ductility.

Few Restrictions. The greatest limitation of deformation processes is the need for a ductile stage in the material. Nearly all metals have ductility at some elevated temperature (the major exception being cast iron) and may be at least hot worked. Working at lower temperature is limited to those materials classed as being ductile.

Deformation Increases Probability of Defects. In metals processing, deformation is fundamental and is successfully performed on virtually every product at some time during the manufacturing cycle. However, improperly controlled, the multitude of manufacturing processes that produce deformation can also produce a multitude of defects. Both manufacturing and NDT personnel must be aware of the capabilities of materials to sustain deformation without the formation of unintended defects. They must also be alert in the early detection of defects caused by deformation and initiate corrective action in the manufacturing process to eliminate the causes.

EFFECTS OF DEFORMATION

WORK HARDENING AND RECRYSTALLIZATION

It has been pointed out in Chapter 4 that when loads which exceed the elastic limit are applied to a metal, a permanent change of position is effected. The properties of the material change because of redistribution of dislocations, change of grain size, and other metallurgical effects. In general, the most pronounced of these changes of property are permanent, and material is said to be *strain hardened*, *cold worked*, or *work hardened*.

Ductility Recoverable. The changes in properties associated with work hardening are due to the strained and unstable position of atoms in the crystalline structure. The changes may be reversed by supplying energy in the form of heat. The atoms, by the process called recrystallization. rearrange themselves into an unstrained condition similar to that which existed before strain hardening. The temperature at which the rearrangement takes place is called the recrystallization temperature and varies with different metals (as shown in Table 11-1).

TABLE 11-1
Recrystallization temperatures for some common metals and alloys

Metals and Alloys		
Material	°C	°F
Aluminum (pure)	80	175
Aluminum alloys	316	600
Copper (pure)	120	250
Copper alloys	316	600
Iron (pure)	400	750
Low carbon steel	540	1000
Magnesium (pure)	65	150
Magnesium alloys	232	450
Zinc	10	50
Tin	−4	25
Lead	−4	25

Ductility Not Lost in Hot Working. When deformation work is performed above the recrystallization temperature, it is termed *hot working* because recrystallization proceeds along with strain hardening. The net result, however, is not different from that which occurs when metal is cold worked, then heated above the recrystallization temperature. Hot working, therefore, permits continuous deformation instead of the cycle of cold working, recrystallizing to regain ductility, and more cold working that would be required for large amounts of deformation below the recrystallization temperature.

EFFECTS OF FLOW RATE

Recrystallization Requires Time. The changes associated with recrystallization depend on finite movements of atoms within the material and on the formation of new grain boundaries, which take finite amounts of time. The actual time required will depend on the relation between the actual temperature and the recrystallization temperature as well as on the rate of straining. However, some critical rate of straining will exist, above which recrystallization cannot proceed fast enough to prevent rupture. If deformation proceeds too rapidly, it is possible, even above the recrystallization temperature, to develop cracks, and the closer the working temperature approaches the recrystallization temperature, the more likely it is for faults of this type to occur.

A different type of strain-rate effect becomes evident at very high (ballistic) speeds; the failure can occur with little plastic flow regardless of the temperature or the ductility a metal may show in a standard tensile test. However, this type of failure is of little concern to processing except in some new special-purpose processes involving high energy rate forming.

DIRECTION EFFECTS

Alignment of Crystals Develops Directional Properties. Any deformation process causes different amounts of plastic flow in different directions. Metals used in manufacturing are ordinarily polycrystalline materials with more or less random orientation of the crystals. In single crystals, a considerable difference in properties along different planes usually exists, but in a polycrystalline metal with random orientation of the crystals, the differences tend to average out. With plastic deformation, crystal fractures, rotations, and reorientation lead to loss of randomness. As a result, the properties become different in different directions.

Directional Effect May Be either Beneficial or Harmful. In products such as drawn wire, this directionality is seldom harmful. The best properties, particularly strength, are developed parallel to the direction of drawing where they are most needed in use. In rolled sheet metal, however, the loss of ductility perpendicular to the direction of rolling but in the plane of the sheet, may cause secondary drawing or bending operations to be difficult or impossible. For some products, the difficulty may be overcome by proper layout of the shape with respect to the direction of rolling, as shown in part A of Figure 11-1. For others, such as shown in part B of Figure 11-1, the part may be oriented 45° with the direction of rolling. Otherwise, the only solution may be recrystallization of the sheet to restore ductility lost not only because of directional effects but also because of cold working. The directionality developed by working is never completely eliminated because even recrystallization grains are likely to have preferred orientations.

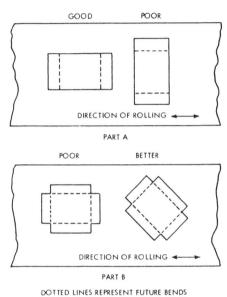

Figure 11-1
Directional effect of rolling on secondary operations

Directional Effects Also from Internal Faults. A second type of directional effect is illustrated in Figure 11-2. Metal as normally cast will contain small quantities of foreign inclusions, such as scale, oxides, and insoluble compounds, and voids or pockets caused by shrinkage and gas evolution during solidification. During working, these defects are elongated in the direction of flow with resulting mechanical property improvement in that direction, generally at the

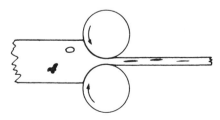

Figure 11-2
Directional effects from elongation of inclusions and voids

expense of properties perpendicular to the direction of flow. Proper design of the product and the tooling, particularly in forging, can take advantage of this directionality, which persists even after heat treatment.

Results of Directionality on NDT. Since internal faults are often flattened and elongated during heavy working, the sensitivity of various NDT methods to detect defects lying in most probable orientations must be considered. For example, it would be of questionable value to radiograph a highly reduced section such as that shown in Figure 11-2 through the short transverse direction. In some cases, grains elongated in one direction can produce markedly different propogation characteristics of ultrasonic energy from one direction to another. The results from other NDT methods can also be more or less critically affected by directionality.

TEMPERATURE AND LOADING SYSTEM EFFECTS

No Theory Fully Explains Plastic Flow. No single theory explains all the phenomena observed in the plastic flow and failure of metals. The following explanation is based on several reasonable assumptions.

Elastic Failure Depends on Shear Stress. Plastic flow occurs only when some critical shear stress is exceeded in the material. This critical shear stress becomes lower as temperature increases except perhaps at temperatures at which recrystallization or crystal transformation take place. Its value also depends on the degree of strain present in the structure, and in the hot-working range, it depends on the rate of deformation. Strain hardening may be interpreted as an increase in the critical stress required for plastic flow.

Fracture Failure Depends on Tensile Stress. Fracture will occur only when some critical tension-stress value is exceeded in the material. This critical tension stress appears to be essentially a constant for a given material and temperature. It drops slightly as the temperature is increased but is not affected by strain hardening.

Deformation Processes Produce Low Tensile Stresses. While the loading system encountered in most deformation processes is quite complicated, the primary loads are usually compressive, and tensile stresses induced are secondary stresses and are often small compared to the compressive and shear stresses. Consequently, much greater percentages of plastic flow may be achieved in an extrusion operation, for example, than can be achieved in a tension test, even below the recrystallization temperature.

GRAIN SIZE

Raw Deformation Material Coarse Grained. For any given metal or alloy, the grain size established on solidification will be determined primarily by the cooling rate. The rate will be determined by the mold material, the superheat present in the liquid metal, the specific heat of the metal, the section thickness of the casting, and the ratio of the metal mass to the mold mass. For most products that are to be used as castings, this ratio is small, and the castings have relatively thin sections. Consequently, a desirably small grain size is established in most castings. However, when it is intended that metal be subjected to some deformation process, it is still necessary to first cast the metal into an ingot. The most desirable forms for ingots are usually quite large with a heavy cross section and a large mass. Therefore, the cooling rate for ingots is quite slow compared to most other castings, and the grain size developed in ingots is very large. For this reason, when ultrasonic testing is called for on ingots and other coarse grained castings, lower frequencies may be necessary. In some austenitic stainless steel castings, even the lowest practical frequencies result in excessive noise from the large grain boundaries. Large grains can also cause diffraction effects in radiographic testing that are undesirable. For best strength and hardness properties for most uses, it is desirable that the grain size be small.

Grain Size Refined Mechanically and Thermally. Any working operation, either hot or cold, results in crystal fractures, rotations, and realignments that produce a small grain size as the material is strain hardened. The actual effect that these grain-size changes have on properties is hard to evaluate, however, because the major property changes are due to the strain hardening. However, if following the strain hardening, recrystallization takes place, either because of subsequent heat treatment if the material was cold worked or during hot working, the grain size immediately after recrystallization will always be small.

Regrowth at Elevated Temperatures. Unfortunately, the small grain size established by recrystallization is not completely fixed, and a metal that is held at too high a temperature or at an elevated temperature for too long a time following recrystallization will undergo the phenomenon of *grain growth* during which the grains will combine and grow to larger size again. Given sufficient freedom (time at elevated temperature) crystals tend to grow to a critical stable size that is dependent mainly on their constituents. Figure 11-3 shows the relations that may exist among grain size, working conditions, and temperature during deformation processes. The slopes of the various lines are only qualitative and may vary, depending on the particular alloys, rates of working, and temperatures; the grain size of the final product will depend on the place where processes stop.

Importance to Nonferrous Materials. The phenomena illustrated by this figure are of extreme importance because, for the majority of nonferrous metals, these are the only methods for grain-size control. For example, if improper heat treatment during recrystallization following cold work has permitted excessive grain growth further cold work would be necessary before grain refinement could be accomplished, and this would not be possible if the final shape had been established. To sum up, grain refinement for metals that exist in only one crystalline form can be accomplished only by hot working, cold working, and recrystallization following strain hardening, and grain growth will occur any time metals are held at excessive temperatures for sufficient time. Various techniques of ultrasonic testing have been developed to semi-quantitatively evaluate grain size.

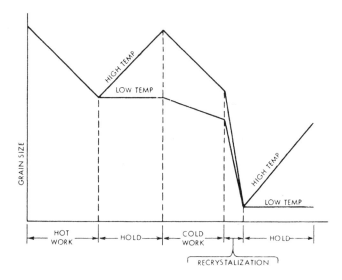

Figure 11-3
Grain size during deformation processing

Ferrous Grain Size Refined Two Ways. For most ferrous alloys, the grain size changes, not only during working and recrystallization, as shown in Figure 11-3, but also at any time ferrous metals are heated through the transformation range. Figure 11-4 shows the nature of these additional changes. For any initial grain size of body-centered cubic iron (below the transformation temperature), the face-centered crystals that form after transformation will always be smaller. However, the size of the face-centered crystals will increase if the metal is held above the transformation temperature; the amount of growth will depend on the temperature level and time.

Best Method by Transformation. Whatever grain size is established in the face-centered crystals will be preserved when the transformation is made back to body-centered cubic iron. This means that the grain size of a steel casting may be refined by heat treatment alone or that the grain size of a hot-worked product that is held at excessively high temperature following working may be refined by heat treatment. Note, however, that this refinement requires that the metal be reheated through the transformation temperature range and also that strain hardening is not a requirement. In general, for ferrous metals, refinement by transformation is much more effective than working or recrystallization.

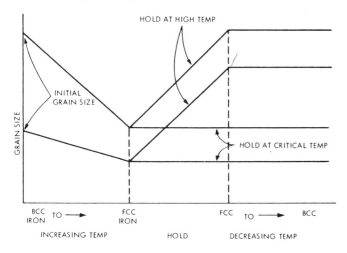

Figure 11-4
Grain-size change with crystal transformation in ferrous metals

RELATIVE EFFECTS OF HOT AND COLD WORKING

MECHANICAL PROPERTIES

Hot-worked Metal Soft and Ductile. Material that has been hot worked will generally exhibit maximum ductility and minimum hardness and strength for a particular composition. Possible exceptions may come from directional effects caused by grain orientation and fibering and effects that cooling from the high temperature may have on the structure of the material. Any effects of strain hardening will have been continuously relieved by recrystallization at the hot-working temperature.

Faults Minimized by Hot Work. Materials that are hot worked start as ingots having relatively large cross sections. As a casting, this shape and size results in pronounced casting defects, such as ingot-type segregation (composition differences within crystals), voids (dendritic microporosity and macroporosity from gas evolution), shrinkage cavities, and inclusion of metallic oxides, slag and other foreign matter. Some of these faults are removed by *cropping* the ingot. Cropping involves the removal and discarding of as much as one-third of the top of the ingot where the largest shrinkage occurs. However, many of the faults still exist in the main body of the ingot but during hot working have their effects minimized by the closing and welding of voids and the elongation of inclusions. Ultrasonic testing is frequently used to locate the optimum plane for cropping. The major voids and inclusions are detected and the ingot is cropped at a location which produces maximum yield with the least defects. The discontinuities that remain in the usable portion of the ingot may have no important effect on the final product, but in some cases may be the origin of a future failure.

Major Deformation by Hot Working. Cold working is used primarily as a finishing process and usually follows hot working that has been used to accomplish the major portion of deformation. The ductility and strength properties of the finished product can be controlled to a rather wide degree by the amount of cold working that is performed in the final stage.

Last Cold Work Effective. With any degree of cold working, the material could be restored to the original conditions of elastic limit and ductility by recrystallization and could then be subjected again to cold deformation. The final strength and ductility will therefore depend on the amount of cold working done after hot working or after the last recrystallization treatment.

Example of Cold Deforming. The application of these principles in practice may be understood by considering the manufacture of cold-finished steel sheet. Nearly all such steel is first hot rolled to a thickness of about 4 millimeters (5/32 inch). If cold finished to 2 millimeter (0.080 inch) thickness by repeated rolling passes with no intermediate heat treatment, the resulting sheet would have high hardness and strength with minimum ductility and be suitable only for products that could be finished with little or no further deformation operations.

If an intermediate anneal were followed by only a few cold-rolling passes, the resulting product would have intermediate hardness, strength, and ductility and be suitable for a limited amount of further cold-working operations, such as shallow drawing or bending with large radii.

If, following the reduction to final thickness, the sheet were annealed, it would have minimum hardness and strength but maximum ductility and would be suitable for deep drawing or other operations involving large amounts of deformation. Any of these further deformation operations would add to the strength and hardness and reduce the ductility.

Reduced Ductility Desirable for Most Machining. When the products of deformation operations are to be further processed by machining, cold finishing is generally desirable, even though the hardness and strength are not needed functionally in the product. The overall machinability of most metals is improved with reduced ductility because of improved finishability. For this reason, much of the bar material to be finished by machining is cold rolled or cold drawn. The compressive stresses left on and near the surface of most cold-worked material are of some benefit when the material is subjected to fatigue conditions in service. Fatigue failures generally start at areas of high tensile stresses on the surface of parts, and the residual surface compressive stress reduces the actual value of surface tensile stress due to applied loads.

FINISH AND ACCURACY

Surface Qualities Affected by High Temperatures. Limitations exist as to the surface finish and accuracy that may be obtained by hot working. Most metals are subject to rapid oxidation at their hot-working temperatures, which are often well above room temperature. In addition to chemical damage, oxide formation is frequently nonuniform, and scale may spall off, exposing new metal to oxygen contact. The surface finish and dimensional accuracy obtainable are largely determined by the rate of oxidation and the tendency for spalling. Such surface conditions can adversely affect the application of some nondestructive tests. Frequently, ultrasonic, penetrant, magnetic particle, and eddy current tests cannot be adequately performed without the removal of rough scale and oxides.

Effects on Low Melting Alloys Not Serious. For aluminum and many other nonferrous alloys, the hot-working temperatures are low enough that oxidation is not serious, and good finishes and close accuracies may be held. For steels, hot-working temperatures are in the range of 950° to 1,300° C where oxidation is rapid. With the scale that forms at these temperatures, it is not possible to obtain good finishes or close dimensions. Tolerances are generally 0.4 millimeter (1/64 inch) or greater on hot-worked steel products. However, cold working of steel can produce finishes limited only by the rollers or dies used in the process and tolerances of 0.025 millimeter are possible.

Decarburization Changes Surface Composition. Steel in particular, because of its high working temperature, is subjected to selective oxidation. The carbon burns at a higher rate than the iron to leave a decarburized shell. Subsequent heat treatment, which depends on carbon content, does not produce the desired results on the surface, and hot-worked steel that is to be hardened by heat treatment needs to have sufficient material removed from the surface to get below the decarburized layer. High carbon hot-rolled steels are usually at least 1/16-inch oversize in the raw-stock stage to permit surface removal. Decarburized layers can severely limit the application of eddy current tests. On the other hand, eddy current techniques can be used under some conditions to provide a measure of the thickness of decarburization, thus assuring adequate removal.

PROCESS REQUIREMENTS

Most cold working is performed at room temperature at which normal variations are unimportant and no specific temperature control is necessary. The increased conduction and radiation rates at elevated temperatures cause control to be much more difficult. In some continuous working processes involving large amounts of deformation, the energy added by the process affects the temperature, and the maintenance of correct temperatures depends on the proper rate of working.

Lower Work Energy Required. In addition to the maintenance of ductility by continuous recrystallization, one of the principal benefits of hot working is that metals are weaker at high temperatures and can be deformed with lower loads and less work. The lower loads result in lighter and more versatile equipment than would be required for equivalent deformation performed cold.

Equipment Life Reduced by High Temperatures. The dies, tools, and other equipment that come into contact with heated materials must be able to maintain adequate hardness and strength. Hot-working tools, therefore, must frequently be made of heat-resisting alloys or be water cooled for satisfactory life. Occasionally, nondestructive tests are used to aid in determining when tooling repairs are needed or when the useful life of the tooling is being approached. Tooling failures can sometimes result in larger scale machine failures, thus warranting the application of NDT as a maintenance procedure.

Millwork, Forging, and Powder Metallurgy 12

Although some of the softer metals that can be found in a relatively ductile condition in nature, such as copper, lead, gold, and silver, have been wrought by hammering methods since the early days in history, most shaping of metal articles in the early days of manufacturing was performed by casting processes. As indicated in earlier chapters, casting is still an important shaping process and is frequently the cheapest and most satisfactory method for producing a useful shape from some materials.

Some Serious Limitations in Casting Processes. Some limitations exist, however, that discourage universal use of cast metal products. Picture, for instance, the problems associated with casting thin sheets of large area in any kind of material. Even with thicknesses of an inch or more, the problems of obtaining uniform thickness and properties over large areas are enormous. Unfortunately, many of the materials that have the best castability have other properties that are unsatisfactory for many applications. Porosity and associated problems reduce the strength. Increased brittleness, leakiness, and poor appearance are faults commonly associated with cast materials.

Deformation Improves Properties. With many metals, the internal structure to provide the best properties can be developed only by deforming the material in the solid state, usually by a process involving cold working. The deformation processes, cold or hot, can often be used to provide the double benefits of property improvement and shape changing at the same time.

Both Ferrous and Nonferrous Metals Deformation Processed. Even with higher costs, the value of improved properties is so great that approximately

80% of the iron-based metals are finish processed as wrought material. Although nearly all metals are and can be cast in the making of some products, a situation similar to that for iron-based metals exists for aluminum-based, copper-based, and other metallic materials, and large percentages of each are deformation worked for improvement of their shapes, dimensions, and properties.

Most Output Requires Further Processing. Most of the output of the mill is in shapes that become raw material for further processing in smaller quantities at some specific user's plant. Typical products of this class include foil for packaging operations, cold-rolled sheet for pressworking operations, bar stock for machined parts, and rough-rolled billets for forging operations.

End Product by Secondary Deformation. The second group of deformation operations are those that are product oriented and are usually performed on a smaller scale in plants fabricating finished products. For practically all of these operations, the raw material is bar or sheet stock that is produced in large quantities as a mill operation. For example, the most convenient raw material for a drop-forging operation might be a 6-inch length of 2-inch-square hot-rolled steel. This would be cut from a long length of 2-inch-square hot-rolled bar. The same 2-inch-square hot-rolled bar might be the most convenient size for other fabricators for forgings, for parts that are to be machined, or for welded assemblies. It is often economical to apply NDT to products intended for secondary operations in order to assure that prior processing defects are not carried forward into secondary processing. The defects might include seams, cracks, and other internal discontinuities of significant size.

Few Mills — Many Fabricators. These smaller fabricators are much greater in number than are producers of mill products. The equipment for the secondary operations is lighter, the first cost of the equipment is generally less, and the total tonnage of metals used by any individual fabricator is small compared to the output of a mill.

MILLWORK
HOT ROLLING

Hot Rolling Is the Common Initial Operation. The chart of Figure 12-1 is typical of steel mills and also applies to most nonferrous mills, although emphasis on the operations will vary for different metals. One of the most common mill operations is the rolling of metal into flat and two dimensionally formed shapes. This is accomplished by passing the material between flat or shaped rollers to set up forces that squeeze the material and cause it to flow to an elongated form while the cross-sectional dimensions are being reduced. For those materials that have little ductility and for large changes of section in any

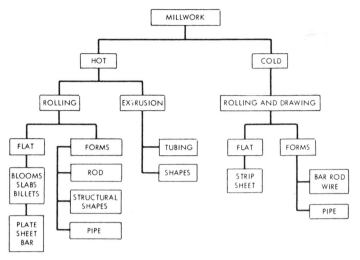

Figure 12-1
Processes and product types of primary mills

material, the work is usually done hot to reduce the energy requirements and to permit ductility recovery by recrystallization as deformation occurs. Some materials must be worked at elevated temperatures in order to attain adequate stability and ductility to preclude fractures during deformation.

Blooms, Slabs, and Billets. Following reduction of the ore or, in the case of steel, following carbon reduction, most materials start as cast ingots that are rolled initially into blooms, slabs, or billets. Blooms and billets are approximately square cross sections of large and small size, respectively, and slabs are rectangular shapes. All are destined for further deformation work by rolling, forging, or extrusion, usually at the same mill but sometimes at an individual fabricator's plant.

Thickness Reduction by Compression. Mill rolling is done by passing the material through rolling stands where rollers, arranged as shown in Figure 12-2, apply

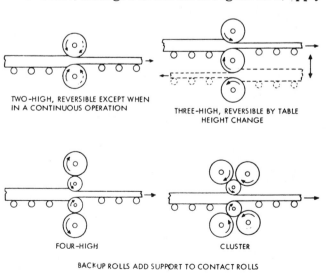

Figure 12-2
Various arrangements of rolls in rolling stands

pressure to reduce the section thickness and elongate the metal. The major portion of stress is compressive and is in such direction that the effect on width dimensions is minor compared to the others.

Blooming Mill Reversible. At the blooming mill where the first deformation work is done on the material, the cast ingot is rolled back and forth between rolls or continuously through sets of stands as the rolls are brought closer together to control the rate of reduction and establish new dimensions. Mechanical manipulators are used to turn the block, or additional vertical rolls are used for making an approximately square cross section bloom or rectangular slab that may be as much as 60 or 70 feet long.

Cast Ingot Defects Removed. As much as one-third of the bloom may be *cropped* (cut away) to eliminate a major portion of the impurities, shrink, and poor quality metal originating in the ingot. Near-surface defects caused by ingot or rolling faults are removed during or following primary rolling by chipping, grinding, or *scarfing* (oxygen torch burning). These long blooms are then sheared to lengths convenient to handle and suitable for the anticipated final material form.

Continuous Casting Eliminates Ingots. Increasing use is being made of continuous casting as a step in steel making. Although the cost of changeover is high, the installation eliminates the making of ingots and their breakdown in the blooming mill. The continuous casting is made in a heavy slab or plate form that can be introduced directly into the hot-roll stands. Another advantage gained is the elimination of ingot cropping.

Billets Smaller Than Blooms. Blooms are frequently reduced to billet size, maximum cross section of 36 square inches, in a similar stand with reversing features, although some installations have been set up with a number of rolling stands in sequence so that billets can be formed by continuous passage through the series.

Hammer Forging for Specials. Some demand exists for small quantities of wrought materials in large shapes not adaptable to rolling. These may be of variable section size, for example, a large steam turbine shaft, or sizes not ordinarily produced by the rolling mill. In these cases, the ingot may be worked to the desired shape by a forging operation, usually between flat-faced hammers.

Continuous Hot Rolling. Following the primary reduction operations in the blooming or slabbing mill, the sections are usually further rolled in some secondary operation, still at the mill. Plate, sheet, and rod shapes are in sufficient demand that many mills produce them in continuous mills. The material proceeds directly from one rolling stand to the next, with progressive reduction and shaping of the cross section and simultaneous elongation along the direction of rolling. Scale-breaking rolls are followed by high pressure water or steam sprays for removal of scale. Both the roughing and finishing operations are done in continuous mills consisting of a number of strands in sequence. Some hot-rolled strip is used directly as it comes from the hot-rolling mill for the making of finished goods such as railway cars, pressure vessels, and boats. Most of the flat hot-rolled steel is further processed by cold rolling.

Surface Oxidation a Problem. As pointed out earlier, the mechanical properties of hot-worked material are affected by the heat to which it is subjected. Working at high temperature permits maximum deformation, but for those materials for which the working temperature is above the oxidation temperature for some of the constituents, burning and scale result, and adverse effects on finish occur. Before use as a product in the hot-rolled state, or before cold-finishing operations are performed, surface cleaning is required. Cleaning is often done by immersing the material in acid baths (pickling) that attack the scale at higher rates than the base metal.

Limited Accuracy in Hot Rolling. Because of differences in working temperatures affecting shrinkage, differences in oxidation depths, and more rapid wear on the rolls, dimensions are more difficult to hold in hot-rolling processes than when finishing is done cold. Tolerances depend to some extent on the shape and the material. For hot-rolled round bars of low carbon steel, they range from ± 0.1 millimeter (0.004 inch) for material up to 10 millimeters (0.4 inch) in diameter to ± 1 millimeter (0.040 inch) for bars 10 centimeters (4 inches) in diameter.

COLD FINISHING

Properties Changed by Cold Working. While most steel is shipped from the mill in the hot-rolled condition, much of the material is cold finished by additional rolling in the cold state or by drawing through dies. The forces set up by either procedure are similar and result in reduction of cross-sectional area. Materials that are treated in this way must have sufficient ductility at the beginning, but that ductility is reduced as the hardness, yield strength, and tensile strength are increased as the deformation progresses.

Flat Products. The flat products of a steel mill are called *strip*, *sheet*, *plate*, or *bar*, depending on the relative widths and thicknesses, and most are cold finished by rolling. For this work the rolling stands are of the four-high type illustrated in Figure 12-3 of the cluster type that performs the same function of permitting small diameter work rolls to be in contact with the material. Figure 12-4 shows typical arrangements of stands for cold-rolling strip or sheet. The tandem mill, with a higher initial investment, is a higher production method but has less flexibility than the single-stand reversing mill. Power for reduction

130 Materials and Processes for NDT Technology

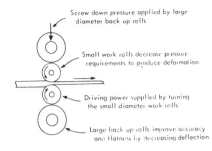

Figure 12-3
Arrangement of conventional four-high rolling stand

may be supplied by the reels alone, by the rolls alone, or by driving both the reels and the rolls. Sheet is normally kept in tension as it passes through the stands.

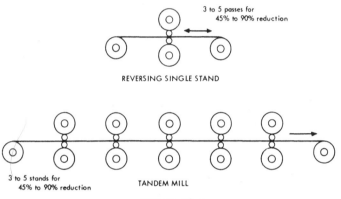

Figure 12-4
Cold reduction methods

Since cold rolled strip and sheet is usually produced with high accuracy thickness requirements, some mills are equipped with on-line ultrasonic or radiation thickness gages. In some sophisticated systems, the output from the thickness gage is fed back to provide roll spacing and tension adjustments while rolling is in process.

A Variety of Bar Shapes Rolled. Bar material can be in the form of square, rectangular, round, hexagonal, and other shapes. In the rolling of strip and sheet, the edges are not confined, and the final width of the sheet may vary. Subsequent to shipping from the mill, the material is normally trimmed to correct width by rotary shears. Most bar shapes are not adaptable to close dimensional control in cold rolling and are therefore finished by drawing through hardened dies. The operation is performed in a machine called a drawbench, shown in Figure 12-5. The end of the oversized hot-rolled bar is first pointed by swaging or forging, then inserted through the die and gripped in the draw head. Connection of the draw-head hook to a moving chain provides the power to draw the material through the die. Reductions generally range from 0.5 to 3 millimeters (1/64 to 1/8 inch). Round stock may also be cold finished by rolling between skewed rollers in a process called *turning* or centerless ground for highest accuracy.

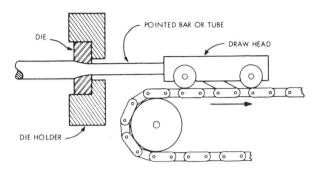

Figure 12-5
Drawbench for cold reduction of bar or tubing

TUBE AND PIPE MAKING

The terms *pipe* and *tube* have no strict distinctions, but in most common use, the term *pipe* refers to a hollow product used to conduct fluids. Except for some relatively thin-walled welded products, tubing is generally seamless.

Pipe and Tubing — Mill Products. Most pipe and tubing products are produced in mills, frequently along with sheet, strip, and bar products. The manufacture of tubular products involves both hot and cold working, in the same order as for other mill products, with hot working being used in the rough forming stages and cold working in the finishing and sizing operations. Most pipe made by welding processes is steel. Some steel and nearly all nonferrous tubular products are made by seamless processes. Regardless of the process, NDT is nearly always used at some stage in the processing of pipe and tubing if the product is to be used in high pressure applications.

Pipe by Welding Bell. One of the oldest but still much used processes for making steel pipe consists of drawing heated bevel-edged *skelp* in lengths of 20 to 40 feet (6 to 12 meters) through a welding bell such as pictured in Figure 12-6. The skelp is gripped by tongs and drawn through the bell where it is formed to tubular shape and the edges pressed together to form a butt-welded joint. Power is supplied by a drawbench as in drawing bar stock.

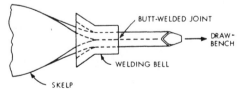

Figure 12-6
Shaping and welding of pipe in a welding bell

Pipe by Roll Welding. Figure 12-7 illustrates the method used for butt welding pipe in a continuous manner. Skelp from a reel passes through a furnace and is drawn through forming rolls where it is shaped. Welding rolls then apply pressure to establish the butt-welded joint. Following the welding station, rollers squeeze the pipe to smaller size after which it

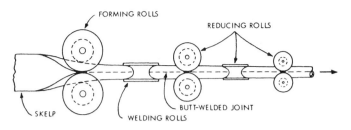

Figure 12-7
Continuous process of butt welding pipe

is cut to length by a flying saw. Both types of butt-welded pipe may require some cold-finishing operations, such as sizing between rollers and straightening by stretching or cross rolling, before being cut to exact length and finished by facing or threading operations. Pipe is produced by pressure butt welding, either short lengths or continuously, in sizes up to 4 inches in nominal diameter.

Resistance Welded Tubing. Light gage steel tubing in sizes up to 40 centimeters (16 inches) in diameter may be produced by resistance welding of stock that has been formed cold by rolls which progressively shape the material from flat strip to tubular form. The general arrangement is shown in Figure 12-8. After forming, the tube passes between electrodes, through which welding current is supplied, and pressure rolls that maintain pressure in the weld area. Because the material is heated only locally, the pressure produces flash on both the inside and outside of the tube. The outside flash is removed by a form cutter immediately following the welding operation. The inside flash may be reduced by a rolling or forging action against a mandrel, depending on size. Because this process uses rolls of strip stock as raw material and is best operated continuously, a flying saw is required to cut the tubing to correct length. Resistance butt welding may be done in a mill, but

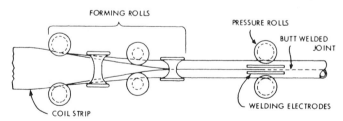

Figure 12-8
Resistance welding of tubing

because of the relatively light equipment needed, it frequently is performed as a secondary operation in a fabricator's plant.

Some Pipe Welded with Filler Metal. For large sizes (from about 15 centimeters [six inches] to an unlimited upper limit) that are needed in relatively small quantities, pipe may be manufactured by forming of plate or sheet and welding by any of the fusion processes. In practice, the *submerged-arc* method,

Millwork, Forging, and Powder Metallurgy 131

which will be discussed in a later chapter, is often the most economical welding procedure. After the edges of the plate have been properly prepared by shearing or machining, the steps shown in Figure 12-9 are followed in forming the pipe.

A relatively small quantity of larger pipe, from about 4 to 75 centimeters (1.5 to 30 inches), in diameter, is lap welded. For this process, the skelp is beveled on the edges as it emerges from the furnace. It is then formed to cylindrical shape with overlapping edges. While at elevated temperature for welding, the tube is passed between a pressure roller and a mandrel for the establishment of welding pressure.

Spiral-welded Pipe. The making of light gage pipe or tubing as pictured in Figure 12-10 can be accomplished by resistance welding of a continuous spiral butt or lap joint. A principal advantage of the process is the light equipment required and the flexibility in changing from one size or one material to another. Any material that can be welded can be fabricated into pipe by this method.

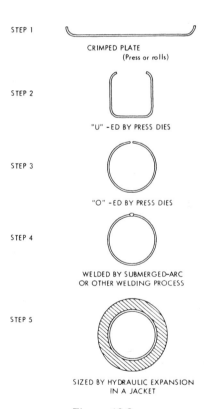

Figure 12-9
Electric welding of large pipe

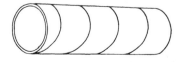

Figure 12-10
Spiral weld pipe

Seamless Tubing. In practice, the term *seamless tubing* refers to a tubular product that is made without welding. The most common method used for steel involves *piercing* of round billets of relatively large cross section and short length, with subsequent deformation operations to control the final diameter, wall thickness, and length. Figure 12-11 shows the most common type of piercing mill used. The skew rollers both flatten and advance the billet with a helical motion. High shear stresses are developed at the center of the billet, at which point the material is forced over a bullet-shaped mandrel.

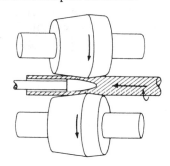

Figure 12-11
Roll piercing of round bar material

Sizing of Seamless Tubing. Subsequent operations include *reeling* and *rotary rolling*, which are similar to piercing and permit the inside diameter to be further enlarged with a reduction of wall thickness. Rolling between grooved rollers reduces both the outside and inside diameter with elongation along the axis of the tube. Much seamless tubing is finished cold by rolling or drawing through dies with the advantages of improved tolerances, surface finish, and mechanical properties. Squares, ovals, and other noncircular shapes may be produced by drawing through special dies and over special mandrels.

Seamless Tubing Useful for Machine Parts. Seamless steel tubing is manufactured from nearly all the common grades of steel, including plain carbon up to 1.5%, AISI alloy steels, and stainless steels of most types. In addition to use for fluid conduction, seamless tubing is also much used as a raw material for many machined parts, such as antifriction bearing races, where considerable material and machine-time savings may be made.

Some Tubing Made by Press Operation. In *cupping* operations, seamless tubing is produced by a press-type operation similar to shell drawing, which will be discussed later. A heated circular disc is forced through a die by a punch to form a closed bottom cylinder. The cylinder may be further processed into a pressure container, or the bottom may be cut off and the tube processed into standard tube types.

NDT of Seamless Tubing. Since the production of seamless tubing can cause tears and other crack-like defects and irregularities in sizing and wall thickness, eddy current testing utilizing encircling coils is frequently applied. By such methods, seamless tubing can be automatically inspected at rates up to several hundred feet per minute.

Perfect Welds Difficult. It is possible to produce welded tubular products that effectively are "seamless." The weld area can have the same properties as the rest of the pipe or tube and may in fact be undetectable after welding. However, this degree of perfection might require heat treatment after welding and additional deformation or machining to produce uniform thickness. In addition, it would be very difficult to produce perfect welds in higher alloy steels, especially in heavy sections. Both radiographic and ultrasonic tests are used for inspection of the welds in pipe produced using a welding process. Fluoroscopic techniques have been widely applied for rapid inspection of the welds. A few ultrasonic systems have been designed to provide pipe weld inspection on-line. Some small-diameter seamless pipes are inspected by eddy current methods that are capable of detecting not only weld defects but defects in the stock material as well.

EXTRUSION

Figure 12-12 shows various extrusion methods. Tubing may be extruded by direct or indirect methods with mandrels as shown. Indirect, or reverse, extrusion requires lower loads but complicates handling of the extruded shape. Lead-sheathed electrical cable is produced by extruding the lead around the cable as it passes through the die.

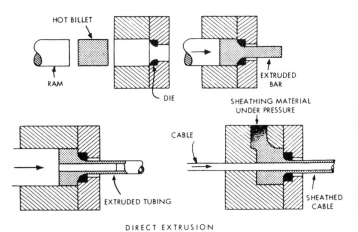

Figure 12-12
Common methods of extrusion

Extrusion a High Energy Process. The high degree of deformation required for extrusion leads to a number of limitations. Most metals are ductile enough for extensive extrusion only at high temperatures. Even then, the loads are very high and require large heavy equipment and large amounts of power. Die materials must be able to withstand the high loads and temperatures without excessive wear. This presents a particularly serious problem with steel, which usually must be heated to about 1,250° C to have sufficient ductility for extrusion.

Steel may be extruded hot with glass as a lubricant, but die life is short; the process is used primarily for steel sections produced in such low quantity that the cost of special rolls could not be justified, and for some high alloy steels that are difficult to forge or roll.

Used Extensively for Nonferrous Materials. The extrusion process is used primarily for forming shapes of aluminum, copper, lead alloys, and plastics. In fact, except for flat stock that may be more economically rolled, extrusion is the principal process used for producing parts having uniform cross sections from these materials. Many metals may be extruded at room temperature. For lead, tin, and zinc, this actually means hot working because the recrystallization temperatures are at or below room temperature, and some heating of the metal occurs as a result of deformation work energy being converted to heat.

Flexible Process but Limited to Uniform Cross Sections. Theoretically, extruded parts have no size restrictions. In practice, the size of the equipment limits the size of the extrusion that can be produced. Dimensional tolerances depend on the material involved, the temperature, and the size of the extrusion. In hot extrusion, the die tends to expand as the material passes through, resulting in a taper to the extruded part. The principal error is in straightness, and most extrusions require straightening. This is accomplished automatically when the extrusion is cold finished by die drawing.

The principal shape limitations are concerned with maintaining uniform cross-sectional thicknesses. Otherwise, the extrusion process is quite flexible; odd and hollow shapes are possible that would be impossible or uneconomical to roll. As previously mentioned, eddy current methods are most commonly applied to testing tubular products that are intended for high pressure applications or high strength structural applications.

FORGING AND ALLIED OPERATIONS

With the exception of some tube-making operations and some cold finish rolling and extrusion, especially on ferrous metals, the operations so far described are all performed almost exclusively in large mills. Mill products usually represent only an intermediate stage of manufacture with no specific finished product in mind. Of the remaining deformation operations, those performed primarily on flat sheet metal will be discussed in a separate chapter.

Forging Is Three Dimensional. In mill operations, the primary shape control is over the uniform cross-sectional shape of a product. In press operations on sheet metal, the thickness of the metal is not directly controlled by the operation. *Forging* operations exhibit three-dimensional control of the shape. For most of these operations, the final shape of the product is forged, and further finishing operations are necessary only because of accuracy limitations of the process.

Forging Dies May Be Open or Closed. The purpose of forging is to confine the metal under sufficient pressure to cause plastic flow. In *open die forging*, the metal is alternately confined in different directions with the final result that three-dimensional control is gained. With *closed impression dies*, the work material is fully confined at least at the completion of the operation in a manner similar to casting except for the state of the material. As in metal mold casting, draft angles are required, and there are similar shape restrictions based on removing the part from the die.

High Compressive Loads Required. The load requirements for forging have led to several means for applying the pressure. In those forging methods in which the metal is worked throughout at the same time, the flow can be produced by constant squeezing pressure or by impact. Because of the large amounts of work energy required and the need to exceed the yield strength throughout the material at the same time, these operations are frequently done hot, and even then the equipment is massive compared to the size of the workpiece, particularly when constant pressure is supplied. For localized flow, the yield strength must generally be exceeded only on small areas at a time, either because of the progressive nature of some rolling-type operations or because of the need to reorient the workpiece periodically to present new areas to be loaded, as in hammer forging or rotary swaging.

NDT OF FORGINGS

Because large volumes of metal are deformed and moved during any forging process, the probability of defect formation can be relatively high. Forgings done at improper temperatures or excessive pressures can exhibit a variety of defects, both surface and sub-surface. Because of the improvement in properties and controlled directionality offered by forgings, they are often used in light-weight critical structures like aircraft and missiles. Even in less demanding applications, forgings are generally selected where high strength and/or directionality is used to advantage.

With the capabilities of NDT to aid in the assurance of high quality, safety, and reliability, forgings are frequently inspected by various methods of NDT. Ultrasonic testing is used principally for detection of internal discontinuities, while magnetic particle and penetrant methods are used for detecting surface flaws. Since many forging defects can be tightly closed and in many cases lie in unexpected orientations due to the large deformations typical in forgings, much care and attention to technique must be applied in NDT of forgings. Forgings often present challenging NDT problems because of the odd shapes and varying cross-sections commonly encountered. Personnel responsible for developing and directing NDT of forgings must have knowledge of the forging process in considerable detail if reliable inspection is expected.

OPEN DIE FORGING

Blacksmithing — A Manual Operation. When the quantity of parts to be manufactured is small and the cost of tooling must be kept low, *blacksmith* or *hammer* forging may be used to alter the shape of the material. One of the simplest examples is the manufacture of a horseshoe from bar stock by using a hammer and anvil with manual power and manipulation. While the village blacksmith is no longer so prevalent, this method still finds wide use industrially for the manufacture of special tools and low quantity products that are often of an experimental nature. Accuracy and shape of the product are greatly dependent on the operator's skill. Because of the close association with the human element, duplication accuracy is limited, and large quantities can seldom be economically produced. The manual operation of blacksmith forging can therefore be used only for relatively light work and is almost always performed hot.

Power Assist for Heavy Work. Hammer forging is an extension of blacksmith forging for larger workpieces in which power is supplied by pneumatic, hydraulic, or mechanical hammers. The operator is still responsible for positioning the work under the hammer but may lay special tools over the hammer faces for producing some shapes. For very heavy workpieces, mechanical supports and handling devices are frequently used as aids.

Rotary Swaging. A rotary swaging machine, as shown in Figure 12-13, is constructed like a straight roller bearing with the inner race replaced by a powered spindle carrying shaped dies in slots. As the spindle rotates, the backs of the dies are forced inward as they pass each roller. Machines of this type are used most frequently for reducing the ends of bar,

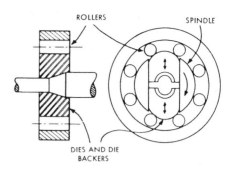

Figure 12-13
Rotary swaging

tube, or wire stock so that it may be started through a die for a drawing operation. Rotary forging may be done either hot or cold, in many cases the choice being determined by the requirements of the drawing operation that follows the forging. In addition to pointing of stock for drawing purposes, the process is used for closing or necking of cylinders and for overall reduction of tubular products.

CLOSED DIE FORGING

Closed Dies Expensive. Most forging was done with flat-faced hammers until just prior to the Civil War when matched metal dies were developed. The process was first used in the production of firearms. With flat-faced hammers and simple grooving tools, no particular connection exists between the tooling and a specific product, and it is feasible to forge even a single part. Matched metal dies, like patterns for castings, must be made for each shape to be forged and become feasible only when the tooling investment can be divided among a sufficiently large number of parts.

Forging and Casting Competitive. To some extent, forging and casting are competitive, even where different materials are involved with each process. As a general rule, the tooling investment is higher for forging than for casting. Thus, the use of forging tends to be restricted to applications in which the higher material properties of steel compared to cast iron or the higher properties of wrought steel compared to cast steel can be made use of in the design. Because forgings compete best in high strength applications, most producers take particular care in raw material selection and inspection. In many cases, either forgings or castings may have adequate properties, and one process has no clear economic advantage over the other.

Material Quality Improved. Proper design for forgings must capitalize on the improvement in properties in certain directions that occurs with metal flow. Voids tend to close and be welded shut under the high heat and pressure, and inclusions are elongated to the degree that they have little effect on the **strength** *in some directions.*

Sequential Steps Necessary. In forging, a suitable quantity of metal is placed or held between the halves of the die while they are open, then forced to conform to the shape of the die by pressure from the dies themselves as they are closed. In drop and press forging, the dies are not completely closed until the forging is completed, with the consequence that, as the dies are closed, the metal may be squeezed to the parting line and be forced out of the die in some places before the closing is completed. To overcome this difficulty, two steps are taken. For most forgings, some preshaping operations are used to insure that approximately the right quantity of metal is already at the proper place in the dies before they are closed. These operations are frequently similar to open die or hammer forging and include *upsetting* (enlarging the cross section by pressure from the end), *drawing* (reducing the cross section of stock throughout), *fullering* (reducing the cross section of stock between the ends), *edging* (distributing the metal to the general contour of the finished stock), and *blocking* (shaping to rough-finished form without detail).

Excess Metal Insures Die Filling. Even with the preshaping operations, it is necessary to provide some excess metal to insure that all parts of the final die cavity are filled. The dies are constructed so that in the closed position a space is left at the parting line through which this excess metal is forced into a *gutter*. The excess metal, called *flash*, is actually part of the forging and must be removed in a secondary operation, generally by trimming in a shearing type of die.

Steel Drop Forged — Nonferrous Materials Press Forged. Theoretically, any metal with enough ductility could be either press forged or drop (impact) forged. In practice, steel is almost exclusively drop forged because of the large capacity presses that would be required for press forging and because the die life would be shortened by the longer time of contact between the die and the heated steel. Most nonferrous metals are press forged. The slow squeezing action in press forging appears to permit deeper flow of the metal than in drop forging, and the dies may have somewhat less draft.

Fast and Accurate but High Setup Cost. Machine forging provides high production rates with little or no material loss and is thus close to an ideal process, providing that tolerances are acceptable, quantities are large enough to cover tooling costs, and the deformation ratios are permissible.

Most common machine forged parts made in very large quantities, such as bolts, rivets, nails, small gear blanks, and great numbers of small automotive fittings, require very little inspection of any kind after the process is in operation. Tool life is long and consistency of product is extremely good. One precaution to be observed is that suitable material continues to be fed to the machines.

FORGING WITH PROGRESSIVE APPLICATION OF PRESSURE

In any closed die forging operation, it is necessary to provide, either by constantly applied pressure or by impact, a great enough load that the compressive strength of the material is exceeded throughout the material for the forging to be completed. Even for forgings of a few pounds, this requires heavy, massive equipment. For a few particular shapes, processes have been developed by which the material is worked only locally with light loads being required, and the area being worked progresses by a rolling action to other parts of the workpiece.

Roll Forging Progressively Reduces Cross Section. Roll forging, illustrated in Figure 12-14, is particularly useful when a cylindrical part is to be elongated throughout part of its length. The drawn section may be tapered, but the process is not capable of upsetting or enlarging the original diameter. In operation, the heated workpiece is placed between the first groove, and the rolls are energized to make one turn, after which the workpiece is moved to the next groove and the operation repeated.

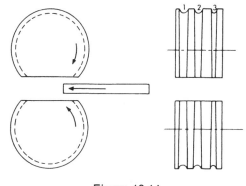

Figure 12-14
Roll forging

POWDER METALLURGY

The definition for the term *powder metallurgy*, as provided by the Committee for Powder Metallurgy of the American Society for Metals, is "The art of producing metal powders and objects shaped from individual, mixed, or alloyed metal powders, with or without the inclusion of nonmetallic constitutents, by pressing or molding objects which may be simultaneously or subsequently heated to produce a coherent mass, either without fusion, or with the fusion of a low melting constituent only."

Originally Developed as a Step in Refining. References to the granulation of gold and silver and subsequent shaping into solid shapes go back as far as 1574. It is also noteworthy that in the nineteenth century more metallic elements were produced in powder form than in any other form. For the most part, however, these were all precious or rare metals for which powder metallurgy was the only practical method of manufacture, and it has only been within the last 50 years that this process has become competitive with more conventional processes in the manufacture of articles from iron, copper, aluminum, and the other more common metals.

Two Unique Advantages. Early developments in powder metallurgy were based on two factors. During the production of platinum, tantalum, osmium, tungsten, and similar refractory metals, reduction was purely a chemical process from which the reduced metal was obtained as a precipitate in flake or powder form. Because furnaces and techniques were not available for complete melting of these materials, the only procedure for producing them in solid form was to press them into coherent masses and *sinter* at temperatures below the melting point. This procedure still applies in the production of some metals, especially tungsten. A second major advantage of the process, which led to early use and is still applied today, is in the production of porous shapes obtained with lighter pressing pressures or lower sintering temperatures. Materials in this form are useful as chemical catalysts, filtering elements, and bearings.

Process Involves a Series of Steps. Figure 12-15 shows the steps ordinarily required in the production of a part by the powder metallurgy process. Suitable powder must first be produced. While theoretically any crystalline material may be fabricated by powder metallurgy, the production of suitable powder has presented restrictions in many cases, either because of difficulty in obtaining adequate purity or because of economic reasons. After selection and blending of the powder and manufacture of a die for the shape to be produced, the powder is pressed to size and shape. The application of heat results in crystalline growth and the production of a homogeneous body.

Little Opportunity for NDT During Processing. Most NDT on powdered metal products is performed after the parts have completed the sintering process. Parts produced from sintered metal powders are nominally inspected as though they were produced by either casting or a deformation process or combination of both. Inclusions, cracks, voids, and density variations can result from improper processing. Since net or near-net shapes are commonly produced by powder metallurgy, NDT is most often called upon to inspect for both surface and sub-surface defects. Radiography is useful to reveal internal voids, cracks, and inclusions and to provide a qualitative assessment of compaction consistency. Conventional penetrant methods are used to detect surface-connected flaws like porosity and cracks.

Properties Influenced by Heat-Pressure Cycle. Various combinations of heat and pressure may be

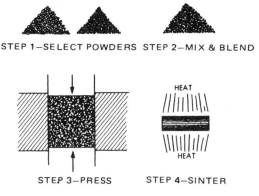

Figure 12-15
Elements of powder metallurgy

used. Some sintering takes place under high pressure at room temperature. However, cold pressing is usually followed by sintering at a temperature somewhat below the lowest melting point of any of the constituents. An intermediate elevated temperature may be used during pressing, then the shape removed from the press and subjected to higher temperature. In hot pressing, the final sintering temperature is applied simultaneously with the pressure.

Mixing Important to Product Quality. Mixing is required for even a single metal powder to promote homogeneity with a random dispersion of particle sizes and shapes. Single materials are often mixed from a variety of sources to develop improved properties. The mixing and blending is even more important for combinations of materials that depend on uniform alloying to develop final properties. Small amounts of organic materials may be added to reduce segregation, and other materials, both organic and inorganic, may be added to act as lubricants during pressing or sometimes in the final product.

PRESSING

Mechanical and Atomic Bonds Established. The bond that is established between particles in powder metallurgy varies all the way from mechanical interlocking to the growing of new, common crystals across the borders of the initial particles. Every atom is surrounded by a force field that is effective at up to a few atom diameters. Proper bonding then depends primarily on bringing adjacent particles close enough together that these atomic forces can be effective. The effective closeness is dependent on both particle size and particle shape. Mixed sizes and shapes, at least with random packing, provide the maximum closeness and the greatest number of contact points.

Deformation Increases Contact Area. Most metals can be plastically deformed, and with these, pressure can be applied to cause the contact points to grow into relatively large areas. The face-centered cubic metals such as nickel, copper, and lead do not work harden readily and can be deformed with comparatively low pressures. The metals that work harden easily and that are also usually harder and stronger to begin with, such as the body-centered cubic structures of iron, tungsten, and vanadium, require much higher pressures to establish suitable contact areas.

High Temperature Accelerates Bonding. Surface atoms will be rearranged both by plastic flow and by mutual attraction with atoms of the adjacent surface. Increasing temperature aids both of these mechanisms by decreasing resistance to plastic flow and by increasing the energy of the atom. Particles that have been severely work hardened as a result of the plastic flow may recrystallize at elevated temperatures, and the new crystals may actually cross the original particle boundary to establish complete atomic bonds.

Multidirectional Forces Desirable. Compacting of metallic powders ideally would be done by applying pressure in all directions at one time. This is usually impractical for commercial use, and most compaction is done along a single axis. Pressure is sometimes applied from one direction only, but in other cases opposing motions are used to reduce the effect of sidewall friction. Figure 12-16 shows the effect of sidewall friction on the density of a compact. The effectiveness of pressing is most often evaluated by measuring the density of the material and expressing

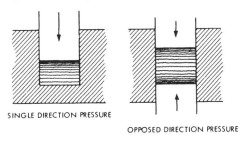

Figure 12-16
Density variation from sidewall friction

it as a percentage of the theoretical density for solid metal of the type being treated. Densities depend on the particle size and shape, the material, the pressure, the time, and the temperature. The figure illustrates the variation in density as the distance from the source of pressure increases. This variation depends primarily on the length to width or diameter ratio of the compact and ranges from as little as 3% for a ratio of one-fourth to as much as 25% for a ratio of two.

Uniform Density Difficult with Complex Shapes. The density variation problem is further complicated by shapes that are other than simple cylinders. Partial solution to this density variation problem may be accomplished by prepressing or the use of multiple punches, as shown in Figure 12-17. Development of pressure by centrifuging may produce more uniform density because each particle of material supplies a force of its own. Rods of various cross-sectional shapes may be extruded with relatively uniform density throughout their length. Thin coatings of powdered materials may be applied to rigid backings by rolling. This procedure is especially useful for various bearing materials.

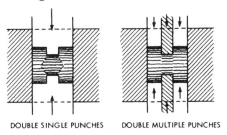

Figure 12-17
Multiple punch for density control

SINTERING

The term *sintering* is used to identify the mechanism by which solid particles are bonded by application of pressure or heat, or both. In its broadest sense, the process includes such procedures as welding, brazing, soldering, firing of ceramics, and union of plastic flakes or granules. Each of the procedures other than those involving metal in powder form are important enough and of such wide usage as to have developed their own language and technology.

Sintering a Nonmelting Procedure. Sintering can be accomplished at room temperature with pressure alone but is most often performed at elevated temperature, either at the same time or after pressure has been applied. With some multiple-constituent compositions, some of the low temperature melting materials may be melted, but in most cases sintering is a fully solid-state process. The two most common sintering procedures are (1) application of heat and pressure together, called *hot pressing*; and (2) application of heat after the particles have been closely packed, by *cold pressing*.

Densities Improved with Hot Pressing. In hot pressing, the plasticity of the particles is greater, and they recrystallize more readily and thus permit high densities to be achieved with lower pressures than would be necessary at lower temperatures. For some materials, densities high enough to provide acceptable properties in the finished product are possible only by hot pressing. However, a number of problems are involved. The high temperatures involved (above 1,370° C for some materials) require expensive die materials whose life may be very short. For some materials, a graphite die is used for each part pressed. Gas that is evolved may be trapped within the material, which leads to porosity defects as in castings.

Protective Atmosphere Desirable. Cold-pressed parts that are subsequently sintered may be heated in conventional manner by being placed in ordinary furnaces or salt baths. In those cases where heat is supplied by convection or radiation, it is usually necessary to provide a protective atmosphere of inert or reducing gas to protect the part from corrosion or chemical change.

SIZING AND POSTSINTERING TREATMENTS

Properties Improved by Deformation. Because of variations of density and other factors, shrinkage of powder metallurgy products during sintering is difficult to control. Parts that require close tolerances must nearly always be finished by some dimensional treatment. Cold working may be used for minor changes of dimensions, but this procedure is limited by the lack of ductility common to powder metallurgy products. Repressing, sometimes referred to as coining, improves the density, strength, and ductility of the material. Even with this process, it is seldom that these properties are equal to those of a similar material produced by fusion. Most commercial deformation working is done by hot working or by cold working with frequent interruptions for recrystallization.

Conventional Heat Treatments Possible. Powder metallurgy products may be heat treated in the same ways as other materials of similar chemical composition, but the treatments are usually not as effective as for the fusion-produced metals, mainly because of the porous structure restricting the heat conductivity. Many of the voids within powder metallurgy products are stress concentration points that not only limit service loads but also increase the stresses arising from thermal gradients during heat treatment. The treatments include resintering for stabilization and homogeneity, annealing for softness, grain refinement for improved ductility, and hardening for improved wear resistance. The hardening processes may be quench hardening of carbon steels, precipitation hardening of nonferrous materials, or surface hardening by carburizing, cyaniding, and nitriding.

Machined When Necessary. The machinability of sintered materials is usually poor, but machining is sometimes necessary to provide final control of dimensions or to establish shapes that are not practical for the powder metallurgy process. With some types of products, such as the cemented carbides, grinding is the common finishing process both to control size and shape and, in many cases, to eliminate the surface produced in the sintering process. The original surfaces may contain faults or inclusions damaging to use of the product.

Properties Improved by Impregnation. One important finishing step is that of impregnation. Inorganic materials, such as oils or waxes, may be impregnated into porous metal products for purposes of lubrication. An entirely different kind of product can be produced by impregnating high melting temperature metals with low melting temperature metals. The principal use of this technique is in the production of *cemented steels*. A porous, skeleton iron compact, which may be produced from low cost iron powder, is impregnated with molten copper. The resulting product has better strength, ductility, and machinability than conventional powder metallurgy parts and may be more readily plated or joined by brazing. Sintered iron has also been impregnated with lead alloys to improve antifriction properties for use as bearings.

Conventional film radiography and fluoroscopy have been effectively utilized on metal/metal impregnations to determine the adequacy of the impregnation. Most often, the material used to impregnate is of much different density than the host compacted material. Unimpregnated voids can be readily seen as can the extent of migration of the impregnating metal.

APPLICATION FOR POWDERED METAL PRODUCTS

Powder metallurgy occupies two rather distinct areas. It is a basic shape-producing method for practically all metals, in direct competition with other methods. In addition, for many refractory (high melting point) materials, both metals and nonmetals, powder metallurgy is the only practical means of shape production. Tungsten is typical of the refractory metals; it has a melting point of 3,400° C, and no satisfactory mold or crucible materials exist for using conventional casting techniques at this temperature. Tantalum and molybdenum are similar. For some other metals, possible to melt, impurities picked up by the liquid from the containers would be undesirable, and powder metallurgy offers the most economical means of obtaining solid shapes.

Cemented Carbides an Important Powder Product. Cemented carbides form one of the most important groups of materials that can be fabricated into solid shapes by powder metallurgy only. The biggest use is for cutting tools and cutting tool tips or inserts, but the cemented carbides are also used for small dies and some applications where wear resistance is important. The principal material used is tungsten carbide, although titanium carbide and tantalum carbide are also used. Some very useful production cutting tools are manufactured by using a strong, tough material as a core and impregnating the surface with titanium carbide or another hard, wear resistant material.

Sintered Bearings. A further area in which powder metallurgy produces products not practical by other means is in the manufacture of materials with controlled low density. One of the first mass-produced powder metallurgy products was sintered porous bronze bearings. After cold pressing, sintering, and sizing, the bearings are impregnated with oil, which in service is made available for lubrication. Although not true fluid film bearings, they provide long service with low maintenance. Porous materials are also useful as filters.

Unusual Alloys Formed by Powder Metallurgy. Composite electrical materials form a group similar to the cemented carbides. Tungsten and other refractory metals in combination with silver, nickel, graphite, or copper find wide application as electrical contacts and commutator brushes; powder metallurgy not only provides a means for producing the combination but also provides the finished shape for the parts. Many of the currently used permanent magnet materials are produced by powder metallurgy.

Powder Metallurgy May Compete Economically with Other Processes. In the applications noted so far, powder metallurgy is in a somewhat noncompetitive position so far as the specific products are concerned. Competition exists between cemented carbides and other cutting tools, but cemented carbides can be fabricated only by powder metallurgy. For many of the other products made of most metals, more direct competition exists between powder metallurgy and other methods strictly as processes where the final products may be identical. In this area, powder metallurgy has a number of advantages and disadvantages. In many cases, the powder metal product is completely finished with no material loss, as a result of the process. Production rates are high; finishes and tolerances are good. Powder metallurgy is particularly useful for shapes with two parallel faces but a complex cylindrical contour in the other dimensions.

Pressworking of Sheet Metal   13

Since its inception about 1850, the working of sheet metal has grown constantly in importance and today is perhaps the most important method of fabricating metal parts. As pointed out in Chapter 12, about 30% of steel mill output is in the form of sheet and plate. Most of this material is further processed by individual fabricators by various pressworking operations that involve deformation, usually cold, and shearing operations in which metal is removed.

Most Metal Consumer Goods — Pressworked. The importance of this form of processing to the economy is especially apparent from an examination of the mass-produced metal consumer goods, such as automobiles, home appliances housings, and office equipment. In addition to exterior housings, many functional parts are made from sheet metal; for typewriters, business machines, and other equipment made in large quantity, the percentages of parts made by this process may approach one hundred.

Ductility Essential. Two requisites to this type of processing are (1) sufficient quantities to justify the high tooling cost that is required and (2) the presence of enough ductility in the material to permit the plastic flow necessary for the particular type of operation being considered. Shearing operations, in which plastic flow is not required, are possible on nearly all sheet materials, even brittle materials such as glass and some plastics. All other pressworking operations are deformation operations, and the degree of processing permissible is dependent on the ductility present in the particular material. Some metals may be cold worked to completion with material as it comes from the mill, some metals require intermediate recrystallization between cold-working operations, and some require heating for more than shearing or minimum deformation operations.

Applied Loads Cause Material Failure. Pressworking operations, whether shearing or deformation, involve the failure of the metal by controlled loading. In shearing operations, the metal is loaded in a manner to cause fracture. In bending, drawing, and other deformation operations, the metal is loaded past the elastic limit to cause plastic flow only, usually by application of tension or bending loads. Unlike most forging operations in which the metal is totally confined, the final thickness of the metal depends on the original thickness and the nature of the operation.

Special Tools — High Cost. The majority of pressworking operations requires special tooling. In most cases, the cutting or forming tools are attached to a standardized *die set* that is mounted in the press. Figure 13-1 shows a simple die set for shearing a round hole or producing a round disc.

Tooling Aligned in Die Set. When mounted in a press, the punch shoe is attached to the ram of the press and the die shoe to the bolster plate, which is the fixed member corresponding to the anvil of a forging press. The guide posts insure proper alignment of the punch and die and simplify the setup because the entire die set may be removed from the press and replaced later without any critical adjustments to be made. In some complex dies, there may be confusion as to which is the die and which is the punch; in normal use, however, the tool member with a recess, hole, or depression is called the *die*, and the *punch* is the member that enters the hole or depression of the die. In most cases, stock feeding and handling problems are simplified by mounting the punch on the top and the die on the bottom of the die set.

Limitations of NDT Applications. Most products using sheet metal seldom require extensive NDT of the sheet metal components. For example, while large quantities of sheet metal are used in aircraft, most of the NDT performed on aircraft during manufacturing is devoted to sub-structure like frames, beams, and spars or heavier structures like landing gear and engine components. There are exceptions to this statement, however, and at some time the NDT specialist is likely to be asked to provide inspection of sheet metal.

While sheet metal inspections are most likely to occur during the service lifetime of the structure to which the sheet metal is integral, some thin metals are used in rockets, some ordnance devices, marine and transportation structures, and pressure vessels. Therefore, some knowledge of the common manufacturing processes for sheet metals will be needed if inspection and NDT is called for. One example of an important application of NDT to thin metals is in thickness control and measurement. Ultrasonic and eddy current methods can both provide highly accurate means for thickness measurement, and are particularly useful where access is limited to only one surface of a thin metal structure. Products made from bending, drawing, spinning, and other forming operations may also require NDT to assure freedom from defects that can result from the large deformations that such operations produce. Ultrasonic, penetrant, and eddy current tests are generally most suitable for detecting tears and cracks that can result from irregularities in the materials or processes used.

SHEARING

Shearing Is a Cutting Operation. The term *shearing*, as used in pressworking, applies specifically to the operation of loading to fracture with opposed edges. *Shear stress* applies to an internal load con-

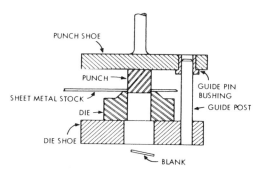

Figure 13-1
Simple die set

dition tending to slide one plane on another, and various amounts of shear stress occur with practically all loading systems. In a *shearing operation*, material is actually loaded by a combination of compressive and bending loads, and the internal stress condition is quite complex. Of real importance is the fact that when the external loads become great enough, the internal stresses will exceed critical values for the material and rupture will occur. The rupture may or may not be preceded by plastic flow, depending on the properties of the particular material.

Shearing Used for a Variety of Purposes. A number of different shearing operations exist with some confusion in names. One of the many ways of classifying these operations is by the process purpose. The purpose may be to produce an external shape, which may either be a finished shape or be the raw material for some other operation; to remove part of the material or cut it in such a way that an opening or indentation is produced; or to remove material that was necessarily left on the part from some other operation. Shearing operations may be grouped as follows:

Stock preparation and blank-producing operations

    Shearing
    Slitting
    Cutoff
    Parting
    Dinking
    Blanking

Hole-making operations
  Punching
  Slotting
  Perforating
  Seminotching
  Notching
  Lancing
  Piercing
Finishing operations
  Trimming
  Shaving

Straight Line Shearing. The term *shearing* generally refers to straight line cutting performed on a squaring shear that has permanently mounted, opposed straight blades. The upper blade is set at an angle to give progressive engagement and reduce the maximum force required. Squaring shears may be used to reduce large sheet or coil stock to smaller size for handling purposes or to produce parts with finished or semifinished shapes, as indicated in Figure 13-2.

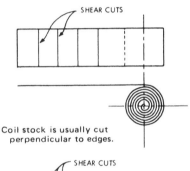

Figure 13-2
Shearing

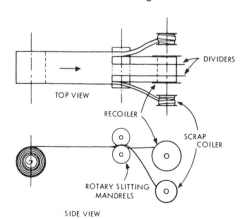

Figure 13-3
Slitting

Slitting. Figure 13-3 shows rotary slitting, which is used primarily for reducing coil stock to narrower widths. Slitting is usually a mill or warehouse operation but occasionally is done by an individual fabricator.

BENDING

In shearing operations, any plastic flow that occurs along the edge is incidental because the purpose of shearing is to cause separation of the metal without any deformation in the sheet itself. *Bending* is intended to cause localized plastic flow about one or more linear axes in the material without causing fracture.

Ductility Required for Bending. Bending is accomplished by loading the material so as to set up stresses that exceed the yield point of the material and cause permanent deformation. Shearing is possible on materials having very low ductility as well as on those having high ductility. Bending is possible only on materials having sufficient ductility to permit the required amount of plastic flow. The severity of bends possible will depend on the ductility. While the degree of bending possible cannot be determined directly from a standard tensile test, this test gives useful comparative data. For two materials, the one showing the greatest percentage of elongation in the tensile test may be bent more severely than the other.

Outside Radius Distorted. Figure 13-4 indicates the nature of the deformation taking place in a bend. The metal on the inside of the radius is subject to high compressive stresses that may cause an increase in width for material that is nearly square in cross section. With any cross section, and regardless of how

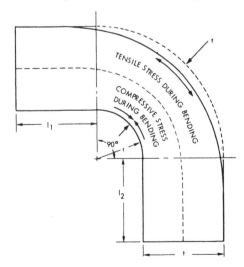

Figure 13-4
Distortion during bending

the operation is performed, the high tensile stresses on the outside of the bend cause thinning of the metal. The degree of thinning will depend on the ratio of bend radius to metal thickness. In practice, the distortion must be considered for two reasons. Unless the metal is actually squeezed at the completion of the bend with sufficient force to cause forging, the outside shape of the bend will not be a true radius and is uncontrolled. On part drawings, the inside radius only should be specified because this radius can be controlled by the tooling.

Forming. By a strict definition, bending would include only operations in which the plastic flow is confined to a narrow straight line region where the bend is made. It is not possible to perform a bend along a curved axis without plastic flow occurring in the material away from the line of the bend. This type of operation would more strictly be called drawing. In practice, however, a number of operations are considered bending that do include some drawing. The term *forming* is sometimes used in a broad sense to include simple bending, multiple bends made along more than one axis, operations that are primarily bending but include some drawing, and some operations that are basically drawing in nature but are of shallow depth or confined to a small area of the workpiece.

Roll Forming — Alternative to Conventional Bending. Roll forming, illustrated in Figure 12-8 in connection with tube making, is not a press operation, but the metal is shaped by means of a continuous bending action. While the completed shape could be produced by bending only, some stretching occurs during the actual forming as the strip changes from flat to formed. Roll forming is used for making tubing, architectural trim, and other similar parts in which a uniform cross section of relatively long length is necessary. The choice between roll forming or shaping by conventional press tooling requires economic analysis. Short parts are frequently made by cutting roll-formed stock to correct length.

DRAWING

Drawing Involves Multiple Stresses. The most complex press operation, from the standpoint of the stresses involved, is *drawing*. In simple bending, a single axis exists about which all the deformation occurs, and the surface area of the material is not significantly altered. Drawing involves not only bending but also stretching and compression of the metal over wide areas. While examples of drawing are many and include such items as automobile fenders and other body parts, aircraft wing and fuselage panels, kitchenware, and square or rectangular box shapes, the simplest illustration is *shell drawing* in which a flat circular blank is pushed through a round die to form a closed-end cup or shell, as shown in Figure 13-5.

In many cases, the dimensions of the required shell are such that it cannot be completed in a single step. A series of dies, (two or more) each smaller in diameter

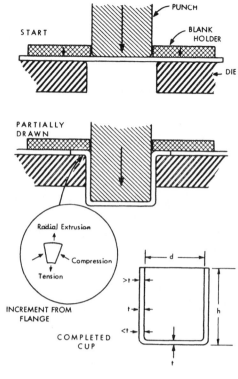

Figure 13-5
Shell drawing

than the previous, are then used to produce the final product dimensions.

Recrystallization May Reduce Number of Steps. An operation might be accomplished with a single redraw if the part were reheated for recrystallization after the first draw to restore the original ductility and permit a greater reduction in the first redrawing operation. The actual choice of a single draw and two redraws as opposed to a single draw, recrystallization, and one redraw would depend on the economics of the particular situation and would involve consideration of quantities, equipment, and other factors.

Single Form Used in Stretch Forming. Figure 13-6 illustrates the short-run method known as *stretch forming*. The sheet to be formed is held under tension with sufficient force to exceed the yield point and pulled down over, or wrapped around, the single form block. Considerable trimming allowance must be left along the edges of the part, and the process is restricted to shallow shapes with no reentrant angles. However, the method is capable of forming operations on large parts and has been used most in the aircraft industry for large wing and body sections.

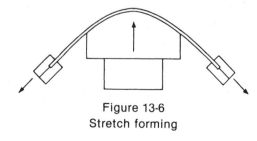

Figure 13-6
Stretch forming

Spinning — Versatile, Low Cost, but Low Quantity, Process. One of the oldest production methods for cylindrical drawn shapes is *spinning*, shown in Figure 13-7. Prior to the manufacture of automobiles and other consumer goods in mass quantities after 1900, spinning was the predominant method for forming deep-drawn shapes and is still used to a considerable extent when low quantities are produced. Most spinning is done cold, but for heavy materials or materials without sufficient ductility at room temperature, elevated temperatures are used. Typical parts include pressure tank ends, kitchenware of a special design and in special metals, and many experimental parts that will, in production, be produced by conventional deep drawing in steel dies.

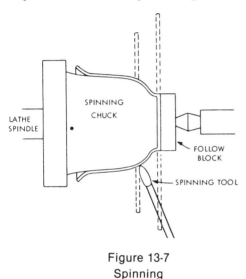

Figure 13-7
Spinning

Tooling is generally low cost and, for light gage ductile materials, wood is the most common form material. Shapes produced may be shallow or deep, and bulging operations are possible with special set-ups. Nearly all metals may be spun, most of them cold. Limitations include the operation time involved and the skill required of the operator because the spinning tool is held and manipulated manually except in highly automatic setups where the process loses its low tooling cost advantage. Usually some thinning of the metal occurs. The problems of wrinkling and tearing are present as in conventional drawing operations, particularly with thinner materials.

NEW DEVELOPMENTS IN SHEET METAL FORMING

Most new developments in this area have at least two features in common. Like the processes just discussed, most are low tooling cost methods, useful for low production quantities, and most make use of a single forming surface instead of matching dies. All of them use nonconventional energy sources, usually some system that releases large amounts of energy in a short time. This feature has led to the use of the term *high energy rate forming* (HERF).

Explosive Forming. Most highly developed of these methods is explosive forming, shown in Figure 13-8. Two general methods have been used. In the first, sheet metal structures are sized or formed by

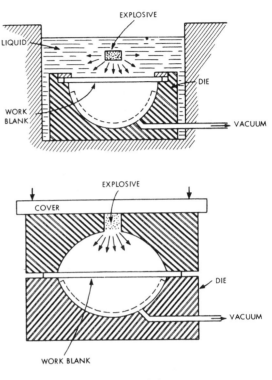

Figure 13-8
Explosive forming

drawing; high explosives detonated in air or in water at some predetermined distance from the workpiece are used. Pressures as high as 4 million psi are developed by the explosion, which creates a shock wave in the fluid medium that transmits the energy to the workpiece. In the second method, a closed die is used, and lower pressures of about 40,000 psi are developed by slower burning propellants or gas mixtures. This system is particularly useful for bulging operations. In either case, a number of advantages exist when the process is compared to conventional press forming. The capital investment is low compared to conventional press equipment, tooling is simple and inexpensive, and sizes can be shaped that would be impractical with conventional equipment; the principal restriction is long production time so that the processes cannot be economically used for quantity production. There has been some indication that greater amounts of deformation may be achieved by explosive forming than by conventional press forming.

Electrical Energy Methods. Similar methods are based on the sudden release of electrical energy stored in banks of condensers. In one method, a spark is created between two electrodes while they are submerged in water or air near the workpiece. In a second method, a high current discharged through a relatively small diameter wire results in vaporization of the wire. In either case, a shock wave is created that transfers energy to the workpiece.

One of the newest methods involves the release of stored electrical energy through a coil near the workpiece, as shown in Figure 13-9. The rapidly created magnetic field induces eddy currents within a conductive (though not necessarily ferromagnetic) workpiece, which sets up fields that interact with the coil fields to create high forces. With properly designed coils, tubular shapes may be expanded into a die or compressed onto a mandrel or various inserts. Flat workpieces may be forced into a shallow drawing die. One of the principal uses has been in assembly of tubular components with end fittings. The system has been called either electromagnetic forming or inductive-repulsive forming. It does not appear to be limited to low production as are most other high energy rate techniques.

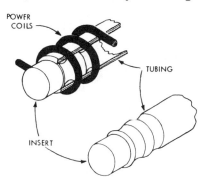

Figure 13-9
Electromagnetic forming

Machining Fundamentals   14

THE MACHINING PROCESS

Machining as a shape-producing method is the most universally used and the most important of all manufacturing processes. Machining is a shape-producing process in which a power-driven device causes material to be removed in *chip* form. Most machining is done with equipment that supports both the workpiece and the *cutting tool*, although in some cases portable equipment is used with unsupported workpieces.

Low Setup Cost for Small Quantities. Machining has two applications in manufacturing. For casting, forging, and pressworking, each specific shape to be produced, even one part, nearly always has a high tooling cost. The shapes that may be produced by welding depend to a large degree on the shapes of raw material that are available. By making use of generally high cost equipment but without special tooling, it is possible, by machining, to start with nearly any form of raw material, so long as the exterior dimensions are great enough, and produce any desired shape from any material. Therefore, machining is usually the preferred method for producing one or a few parts, even when the design of the part would logically lead to casting, forging, or pressworking if a high quantity were to be produced.

Close Accuracies, Good Finishes. The second application for machining is based on the high accuracies and surface finishes possible. Many of the parts machined in low quantities would be produced with lower but acceptable tolerances if produced in high quantities by some other process. On the other hand, many parts are given their general shapes by some high quantity deformation process and machined only on selected surfaces where high accuracies are needed. Internal threads, for example, are seldom produced by any means other than machining, and small holes in pressworked parts are machined following the pressworking operations.

Tool Applies Controlled Loading to Cause Material Failure. Machining, as well as forging and pressworking, is based on the fact that one material can be harder and stronger than another. If the harder one is properly shaped, it can be called a tool; when the tool is brought into contact with a weaker workpiece with sufficient force, failure results in the workpiece. All deformation operations are based on the proper control of this failure. The loading is controlled in machining so as to produce only localized failure in the workpiece, which results in the removal of material in the form of chips without significant deformation in other parts of the workpiece.

Processes Differ Primarily in Energy Use. To understand better what is involved in machining, it might be well to consider what is involved in some of the other fabrication processes and then see how machining differs.

Casting—Heat Energy. In casting, energy is added in the form of heat so that the internal structure of the metal is changed and it becomes liquid. In this state, the metal is forced by pressure, which may consist of only the force of gravity, into a shaped cavity where it is allowed to solidify. The shape changing is therefore accomplished with the metal in such condition that the energy form is primarily that of heat, and little energy in the form of force is required.

Welding—Heat and Force Energy. Welding involves placing the metal in a molten or near-molten condition, again by the addition of heat, and affecting a union by fusion, which may involve pressure. Neither of these processes changes the shape of the metal while it is in its solid and strong state.

Deformation Processes—Mainly Force Energy over Large Areas. In forging, bending, drawing, rolling and extruding operations, advantage is taken of the property of metals to deform plastically. In forging, rolling, and extrusion, pressure loading is applied so that the primary stresses produced in the metal are compression. In drawing operations, metal is pulled or drawn through a controlling die with a complex stress distribution involving tension and compression at the point of metal flow. The forces used to produce shapes by bending result in compressive stresses on one side of the material and tensile stresses on the other. All of these operations are basically the same in the sense that a given quantity of metal is placed in a new shape without any appreciable change in volume.

Machining—Localized Force Energy. To shape a product by material removal in machining, a fracture failure must be caused at the desired location. Loading of the material by relative motion of the tool causes plastic deformation of the material both before and after the chip formation. All materials, however brittle they may seem, undergo some plastic deformation in the machining process. In machining, the energy is in the form of a localized force that causes plastic deformation and fracture to produce a chip.

CHIP FORMATION

Some controversy exists over the theory that best explains the formation of a chip in metal cutting. The following, whether or not it is completely correct, is one of the more generally believed theories that serves a good purpose in helping provide a better understanding for tool design and use.

The Tool Is Simply a Loading Device. First, let it be understood that a cutting tool is merely a device for applying external loads to the work material. If a tool is strong enough that it will not fail and the work is rigid enough to resist deflection away from the tool, a chip will be produced by a relative motion between the two, regardless of the shape of the cutting tool edge in contact with the work. Although any shape of edge may cause a chip to be formed, certain shapes will be more efficient in use of work energy than others, and will exhibit less tendency to set up forces of such magnitude that the tool or work will be damaged.

Forces Are Created by Tool Motion. Figure 14-1 shows a single-point tool moving into the work and subjecting it to compressive loading. The load may be broken down into two forces: a force perpendicular to the tool *face*, which is called the normal force; and

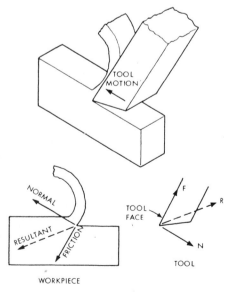

Figure 14-1
Forces in chip formation

because this is a dynamic situation, a force along the tool face, which is the friction force. The two forces may be added vectorially to produce a resultant that, as is shown, projects downward into the work material. The direction and magnitude of the resultant are dependent on its two component forces and are influenced by the angle of the tool face and the coefficient of friction between the chip and tool face. Equal and opposite forces will occur in the tool, but these are of little interest, providing the tool is strong enough to withstand the applied loads.

Stresses Cause Material Failure. As pointed out in the discussion of stresses in Chapter 3, an external force applied in a single direction may set up stresses in other directions within the material. Figure 14-2 shows that maximum shear stresses are induced at an angle of approximately 45° to the direction of the resultant and that the plane region extending from

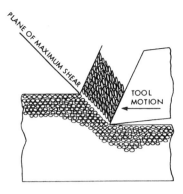

Figure 14-3
Deformation of chip material

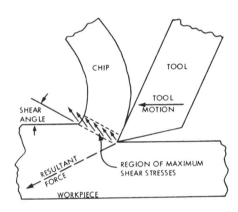

Figure 14-2
Shear stresses in chip formation

the tip of the cutting tool to the uncut surface of the work is subjected to these maximum shear stresses. In Chapter 11, it was indicated that plastic flow will occur when the shear stresses reach a critical value for any material. As plastic flow occurs along this plane, work hardening will increase resistance to further flow, higher stresses will develop, and fracture failure near the tip of the tool will cause the separation of a chip that will ride over the face of the tool and thereby create the friction which causes one of the component forces acting on the work.

Chip Form Dependent on Material and Force Direction. If the material is of brittle nature, it will be able to stand only a small amount of plastic deformation without fracture failure. If it is of ductile nature, the chip may hold together in a long continuous strip or ribbon, deforming considerably, but not fracturing except near the tool tip where it separates from the parent stock.

Figure 14-3 indicates the probable nature of the deformation in the chip, assuming a homogeneous work material with uniform round crystals. Because actual materials are not completely homogeneous, a single plane of maximum shear probably does not exist, but rather there is a shifting plane creating a region or zone in which plastic flow of the work material occurs. In this region, the material is deformed in such a way that the chip is always thicker and shorter than the material from which it is made. The amount of change in shape is dependent not only on the characteristics of the work material but also on the direction of the applied forces.

Chip Types. There are three distinct types of chips that are produced in machining depending primarily upon the machining qualities of the work material but also influenced by tool shape, cutting speed, and other factors.

With brittle materials the chips universally break into segments because of the inability of these materials to withstand the deformation of chip formation without fracture. Tool shape and use to produce chips of small pitch (short segments) usually produces best results concerning tool life and surface finish.

When ductile materials are machined the resulting chips tend to hold together producing chips that are continuous or of relatively long length before breaking free. Ideally all the material that breaks away from the base material will escape uniformly and continuously over the tool face leaving a smooth work surface that has been disturbed to only a minimum degree.

Unfortunately most chips from ductile materials tend to form somewhat intermittently with some material adhering to the tip temporarily, then escaping both over and under the tool tip. This leads to fluctuating forces which may cause chatter and leaves partially removed particles on the work surface affecting the finish and wear qualities.

In addition to the above three identifiable types of chips, under many machining conditions chips may have varying degrees of the qualities of each and cannot be categorized as a single type.

Surface Effects From Machining. Regardless of the type chip produced during machining, force must be used and energy expended resulting in material deformation and heat formation.

The force required to form the chip is in such direction, as indicated in Figure 14-1 and 14-2, that it not only deforms the material of the chip but also applies high pressure to the newly created work surface that passes under the tool. With some materials this deformation action may result in fine surface cracks.

Although machining is not normally a heat dependent process, evidence of its presence is usually quite clear. Immediately after machining a part will feel warm, or hot, depending on the amount of material removed or there may even be considerable radiant heat from the part or chips. In many chips, red heat can be observed at the tool tip as it cuts, and in nearly all cases chips will show discoloration from being exposed to air at room temperature. Except for certain nonferrous materials, grinding displays sparks of burning materials as a result of cutting action.

In many cases such as grinding or other high cutting speed operations, very high localized temperatures approaching the melting temperature of the work material may be generated. High temperature gradients can set up thermal stresses sufficient to cause small surface cracks that could be harmful. For critical parts inspection by NDT may be required to detect these defects and determine their frequency. It should be noted that these defects are often disguised by *smear* metal wiped over the surface by the machining operation—even during some of the finest grinding work.

CUTTING TOOL MATERIALS

Tool materials have always played an important part in the economy of the world. In the earliest days of history, stone was the principal tool material. As late as the nineteenth century, the American Indian used flint for arrow points, spear heads, knives, and other types of cutting edges. Even today some primitive peoples use stone as one of the main tool materials. During the Bronze Age, copper alloys took the place of stone in the more civilized areas. With the discovery of iron and steel, a tool material was found that has been used for hundreds of years and was added to only after the Industrial Revolution and the development of mass-production principles called for tool materials that could operate at higher speeds. Since the beginning of the twentieth century, a number of new tool materials have been developed, and most of them play some part in current manufacturing.

Strength at Elevated Temperatures—an Important Characteristic. The requirements for a satisfactory cutting tool material are that it be harder and stronger than the material it is to cut, that it be abrasion resistant to reduce wear, and that it be able to maintain these properties at the temperatures to which it will be exposed when cutting. The latter requirement has become increasingly important during recent years because of the development of work materials with superior properties and the need for operating at higher cutting speeds to increase production. The principal difference between the tool materials in common use is in their ability to maintain hardness and strength at elevated temperatures. Some of the tool materials with their principal characteristics are as follows:

Carbon Tool Steel. A plain high carbon steel containing from 0.9% to 1.2% carbon. Machinable in its annealed condition. Heat treat hardened and tempered after machined or forged to shape. Little used as a cutting tool material except for some special low use tools.

High Speed Steel (HSS). An alloy steel that maintains cutting hardness and strength to about 550°C (1000°F), approximately twice that of carbon tool steel. Used for many drills, reamers, milling cutters, and other cutting tools where the cutting speed has relatively small effect on the overall manufacturing cost.

Cast Nonferrous Alloys. Alloys that are not normally machinable except by grinding. As a cutting tool it is used to some degree for machining cast iron and malleable iron because of its high abrasion resistance. More commonly used as a structural material or coating because of its chemical and abrasion resistance.

Cemented Carbides. A powder metallurgy product of tungsten, titanium, and/or tantalum carbides combined in various mixtures with cobalt or nickel to produce a variety of hardness and strength properties. The single most important industrial cutting tool group, in present day manufacturing. Used most as a cutting tool tip or insert. Withstands temperatures over 1100°C (2000°F).

Ceramics or Cermets. Another powder metallurgy product, the most successful of which has been made of almost pure aluminum oxide. Less shock resistant than most of the cemented carbides but economical to use for removal of large amounts of material with uninterrupted cuts or for machining some hard materials that would otherwise require grinding.

Diamonds. The hardest material known to man but brittle and subject to failure from thermal shock. Used in single crystal or sintered polycrystal form for machining low tensile strength materials (aluminum, sintered bronze, graphite, and some plastics) with high speed, shallow cuts producing hard quality finishes.

Coated Tool Materials. Strong, shock resistant tool bodies coated with hard, wear resistant materials. An example is titanium carbide impregnated into the surface of high speed steel to take advantage of the values of each.

ABRASIVES

The above mentioned tool materials are used for single point tools or for multipoint tools in which the cutting edges are carefully related to each other. Another group of materials known as *abrasive* are used as wheels, sticks, or stones, or in free form. In use each abrasive grain as it makes contact with the work cuts by exactly the same mechanism as would a single point cutting tool. The random shape of the grains together with their random orientation creates a multitude of cutting conditions which continually vary as tool wear occurs.

Aluminum Oxide. A hard strong grain, much larger than when used in a ceramic cutting tool, used for the vast majority of grinding tools and applications.

Silicon Oxide. Harder and sharper grains than aluminum oxide but more brittle so they break easier in use. Used largely for tool grinding work and for grinding low strength materials.

Diamond. The same material used for single point tools but in this case crushed, graded, and usually supported by a metal or ceramic back up material. Used to a great extent for finish grinding some of the harder cutting tools.

Boron Cubic Nitride. A relative newcomer (1969). This material approaches the hardness of the diamond. It has had some success as a lapping material and shows promise in wheels for tool grinding.

MACHINE TOOLS

Although there are many kinds of machines used in manufacturing industry, the term *machine tools* has been assigned to that group of equipment designed to hold a cutting tool and a workpiece and establish a suitable set of motions between them to remove material from the work in chip form. There are two relative motions necessary for a controlled surface to be established. One is the cutting motion which supplies the power for chip forming. The other motion, or sometimes motions, is the feed motion which presents new material to the cutting edge and in combination with the cutting motion establishes the shape being cut. The common available combination of motions is shown in Figure 14-4.

There are five basic types of machine tools that differ in the combination of cutting and feed motions they permit and in the usual kind of cutting tool for which they are designed. Typical machine tools are illustrated in Figures 14-5 through 14-12.

Turning and Boring. These machines normally rotate the workpiece to produce the cutting motion and feed a single point tool parallel to the work axis or at some angle to it. External cylindrical machining is called *turning*, internal cylindrical machining is called *boring*, and making a flat surface by feeding the tool perpendicular to the axis of revolution is termed *facing*.

Drilling. A special fluted tool with two or more cutting lips on its exposed end is called a drill and is rotated and advanced axially into the workpiece by use of a *drill press*. The principal work is the making of, or enlarging of, cylindrical holes.

Milling. There are a great variety of *milling machines* which like the drill press employ special multi-edge cutters. Except for some special production type milling machines, this equipment premits multidirection feeding and the cutters perform their principal cutting on their periphery edges.

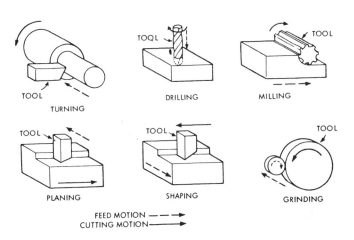

Figure 14-4
Feed and cutting motions

Figure 14-5
Small size engine lathe. Versatile tool-room machine involving a large human element. Seldom used for production of comsumer goods.

Straight Line Machines. One group of machine tools provide straight line cutting motion for its cutting action. This includes the shaper (straight line motion of the cutter), the planer (straight line motion of the workpiece), and the broach (straight line motion of a special multitooth cutter). Because of the high cost of the special cutter, broaching is used only for production quantity machining but the shaper and planer are job-shop type machines.

152 Materials and Processes for NDT Technology

Figure 14-6
Turret lathe. Economical for multiple production even in small quantities because of setup simplicity. Tools index into position as needed from the tailstock turret. Additional tools are mounted on the turret tool post on the side carriage.

Figure 14-8
Horizontal crank shaper. A tool-room machine with great versatility and simple setup but relatively slow productivity

Figure 14-7
Band cutoff saw

Figure 14-9
A numerically controlled machining center. A point-to-point two-dimensional machine with multiple depth stops and manual tool change. Suitable for combinations of drilling, reaming, boring, tapping, and some straight line milling.

Grinding. Because any shape surface made by any other process or machine may require grinding as a finishing operation, there are a great number of grinding machine types. The machine drive rotating abrasive wheels at high cutting speed for the cutting motion and usually produce multiple feed motions simultaneously so the wheel contact may cover the desired surface.

Production Equipment. The machine tool types are constructed from those that require complete atten-

Machining Fundamentals 153

Figure 14-10
A six-spindle drill press that can be set up with
a variety of tools for a series of sequential operations

Figure 14-12
Plain vertical milling machine. Shown fitted with
shell end-milling cutter and fixtures designed for
job-shop production.

Figure 14-11
A flat surface broaching cutter mounted
on the ram of a vertical broaching machine.
Each cutter tooth projects above the preceding ones,
and a surface is completed in a single pass.

tion and considerable operating skill from the operator to production types that are fully automatic. Companies manufacturing large quantities of like product obtain their greatest economy using transfer type machines connected together with automatic handling systems to move the product from one station to the next. These machines frequently include built-in output during continuous manufacturing.

MACHINABILITY

Nearly everyone has at some time used a pocket knife to whittle some shape from wood. While such an operation does not fill all the needs of a machining definition, it is nevertheless a chip-forming operation that uses a hard and strong tool to cause localized failure in a workpiece. The whittler has doubtlessly also noticed that some woods are easier to shape than others. He is faced with an inherent difference in the "whittleability" of different kinds of wood. This ease of working is affected not only by the kind of wood but also by the moisture content and the state of seasoning.

A similar consideration arises in machining metals. Different metals may be cut at different rates, different amounts of power are required, and different finishes are obtained. These differences depend not only on the kind of metal or alloy but also on its prior history of processing, including deformation and heat-treating operations that affect its hardness, strength, and grain structure.

Machinability—an Inherent Material Quality. The term *machinability* is used to describe the relative ease with which any material may be machined. In one respect, the term is like the word *strength*, for a material can have tensile strength, shear strength, impact strength, fatigue strength, and compressive strength, all of which are measured in different ways and any one of which does not necessarily correlate with the others. That is, materials having equal tensile strengths do not always have the same impact strength or fatigue strength. Three different measurements—finish, power consumption, and tool life—may be considered in machinability. Unlike measurements of strength properties, these do not always give precise numerical information, but are more often relative to some standard.

FINISH

To have real meaning, any measurement of finish would have to be made with all the variables that might affect finish under strict control, and the values obtained would be reliable only for a particular set of machining conditions. The relative finishability of different materials has somewhat more reliability. For example, brass normally finishes better than steel under any given set of conditions.

Waviness—Broad Uniform Variations. The geometry of any surface is affected to different degrees by different factors. The gross conformance of a surface to its intended or theoretical shape is controlled by the accuracy of the machine tool motions, by vibrations or deflections of the machine tool or workpiece, and by deformations that may occur as the result of temperature change or the release of residual stresses. The term *waviness* is used to describe those variations of conformance that are relatively widely spaced or large in size.

Roughness—Fine Uniform Variations. The term *roughness* is used to refer to the relatively finely spaced surface irregularities, the height, width, and direction of which establish the predominant surface pattern. These irregularities are superimposed on the waviness. Roughness may be due to higher frequency vibrations, to feed marks occurring as a result of the combination of tool shape and machine tool relative motions, or to the particles of built-up edge that have escaped under the cutting edge and been smeared on the finished surface.

Lay—Direction of Tool Mark Pattern. The *lay* of a surface is the direction of the predominant surface pattern. Lay is determined primarily by the direction of the cutting motion used to machine the surface and may be single direction, circular, or random in nature.

The exact classification of many surface irregularities frequently depends on the method of measurement. Most surface-finish measuring instruments may be adjusted to respond only to variations of less than some particular width so that feed marks, low frequency vibration, or chatter may or may not be recorded in the measurement. Measurements of both waviness and surface roughness will generally be different when measured in different direction because of the effect of the lay.

Imperfections Usually Random. Any surface may contain, in addition to roughness and waviness, randomly distributed flaws or imperfections. These are most often due to inherent faults, such as inclusions or voids in the material, that are exposed only when the outside surface is machined away. Scratches or marks caused by mishandling also fall in this category.

Finish Not Always Predictable. While surface finish depends on many variables and in many cases on the particular combination of all the variables, especially when vibration is encountered, it is possible to make some general statements about the effect of the more important factors. Table 14-1 shows the most likely effect on surface finish caused by increasing the more important machining variables from some standard set of conditions. The predicted results are intended to be qualitative only and even then apply only if one variable at a time is changed.

TABLE 14-1
Relation of machining variables to surface finish

Variable	Finish Effect with Increase of Variable
Cutting speed	Improvement
Feed	Deterioration (degree dependent on nose shape)
Depth of cut	Deterioration
True rake angle	Improvement
Relief angle	Little effect
Nose radius	Improvement
Work hardness	Improvement

There are major exceptions when vibration is considered. Changing almost any condition can often stop vibration, even when the change is in the direction that would otherwise produce a poorer finish. Further exceptions occur at feed rates and depths of cut near zero. With either of these variables at very low values, finish is frequently poor, especially as tools become dull. With a very small depth of cut or feed and a worn tool, the rake angle is decreased with increased forces and greater tendency for built-up edge.

Some compromise is frequently involved among finish, tool life, and machining time. Decreasing the depth of cut or feed may improve finish, but either change would increase machining time. Increasing cutting speed almost universally decreases tool life. Increasing the rake angle may make the tool subject to edge chipping or fracture failure or may induce chatter.

NUMERICAL CONTROL

Numerical control (N/C) systems are auxiliary machine control equipment that may be applied to almost any kind of mechanical device which can function by repeating a certain cycle of operation. This relatively new development is especially important in the manufacturing field because it can be applied to most machine tool types and some other machine equipment such as punches, welding equipment, cutting torches, and even drafting machines.

Greatest Value for Small to Medium Quantities. Although it would be possible to retrofit a standard machine with N/C, the results obtained would be very limited in scope, accuracy, and time saving so that only rarely would such action be economically justified. Practically all N/C equipment is of special design with an integrated control system such that the total cost may be many times that of a conventional machine designed to perform similar product work. Because the cost is high it seems unusual that is it most economical to use N/C equipment on relatively small quantity lots, only occasionally exceeding one or two hundred pieces.

The major benefits received from N/C include reduction of the human element relation to the product with resulting improvement of consistency requiring less inspection. The reason for its value in small lot sizes is based on the short set up time, particularly when the program has already been prepared for previous runs. The equipment can therefore be shifted from one product part to another by changing the tape and available tools with very little time loss.

At the present time, and likely into the indefinite future, large quantity manufacturing of the continuous type can be done most inexpensively with specialized, single purpose machines, usually tied together with mechanical handling equipment, and in many cases including most of the inspection equipment needed to maintain quality.

Principles of Operation. Numerical control consists of storing information in the form of numbers and supplying that information in proper order to the machine to cause the machine to go through some predetermined cycle of operation. Some machines are of conventional design and may be operated manually as well as by N/C. Others are so special that manual control is very difficult and in order to exhibit their greatest value may need to use a computer generated program.

Program Storage. Several storage media such as magnetic tape, punched cards, and others can and have been used, but currently industry has generally accepted a 1-inch wide 8 channel tape as a standard input medium. Figure 14-13 shows a short section of such a tape displaying two words of information to describe X and Y axis positions. The presence or absence of holes at various locations along the eight channels are bits of information that make up characters and words which can be interpreted by the machine reader to initiate action.

Readers Usually of One of Two Types. Some readers are of mechanical type constructed with spring loaded pins that can complete an electrical circuit wherever a hole exists to permit electrical contact. This type reader usually reads a block of information at a time, actuating relays or other electrical devices in the control system.

Other readers are photoelectric and usually read only one character (line) at a time but do this so rapidly that the tape is in continuous motion and the information is recorded.

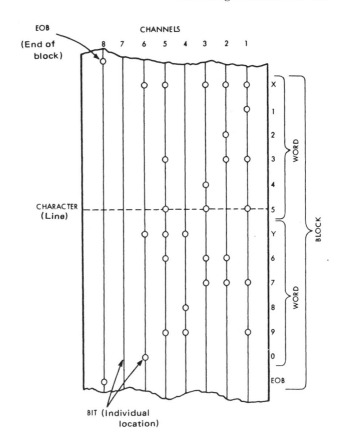

Figure 14-13
Section of N/C tape

Some machines use the supplied information immediately, but others operate more smoothly by reading ahead of the action to a greater degree and storing the information until it is needed.

Most Machines of Closed Loop Design. A small number of machines have been designed to obey their commands without response to the control system. Most though are designed with transducers in the machine elements which generate feedback signals for the control system. As long as error exists between the comparer feedback and command signals, movement continues. As soon as the comparison error disappears, the next command takes over.

Controlled Motions Called Axes. Some machines such as a simple N/C drill which moves the work only through an "x" and "y" axis under a spindle that moves only with a single position stop are called two axis machines. Such is shown in Figure 14-14. Others are more complex with a "z" axis (spindle control) and sometimes rotational motion of the work about one or more axes to present different faces of the work to the cutting tool.

Many machines are also constructed with multiple tool holding racks and the capability of selecting and using particular coded tools as called for by the program on the tape.

Some machines move from one point to another with no control over the path traveled to arrive at the new position. Others differ from the simpler "point to

Figure 14-14
Numerically controlled drilling and milling machine

Figure 14-15
A continuous path numerically controlled lathe. tape reader and controls are shown at the left.

point" machine by operating through a continuously controlled path permitting the generation of accurate curves and shapes. Figure 14-15 pictures an N/C lathe of this type.

Machine Types. Most of the machine tools fitted with N/C fall in the general categories of lathes, drilling machines, and milling machines although many are combinations of the drilling and milling types and have been given the general name of *machining center* because of the great variety of work that can be accomplished on a workpiece in a single set up.

N/C Advantages. Reduced tooling costs by use of simplified jigs and fixtures. Low setup time and cost (most important values). Excellent repeatability with relatively good accuracy. Fewer errors from human fallibility. Reduced lead time.

N/C Disadvantages. Original machine more costly. Machines more complex thus requiring more maintenance. Operators and maintenance personnel require special training and skills. Machines usually require more floor space than conventional types. Effective use necessitates coordination of design with the equipment.

Computers. Many programs prepared for N/C can best be done by use of a computer to perform lengthy computations and turn out a tape ready for use. In addition to computer assisted programming, some equipment is designed to be directly operated by a computer which uses its own memory bank for program storage and eliminates the need for a tape and tape reader. The N/C equipment may be connected directly to its own small computer or may be included in a bank of machines controlled by a large computer. In either case, the program can normally be edited or corrected at the computer keyboard or can be quickly shifted to an entirely new program.

Miscellaneous Processes 15

The processes that have been discussed in previous chapters have all fit the conventional definitions for casting (melt and flow), deformation (plastic flow in the solid state), welding (bonds formed by heat or pressure or both), or machining (chip formation by a cutting tool). In some respects the processing of plastics follows these same conventional methods, but differences in their structure and properties from those of metallic materials causes different processing problems and prevents their being treated in exactly the same ways.

Adhesive bonding, although closely related to welding as a joining process, is also a somewhat special process that seems to fit better with this miscellaneous group.

A number of processes for shaping metallic materials do not fit the standard categories of metal processing. Most are relatively new processes that are still undergoing development. Most are of importance primarily for some special purpose and do not compete economically with the more conventional processes on a wide scale. Several of them have been developed largely because of the need to shape new high strength and temperature-resistant alloys that are not easily worked by the older processes.

PLASTIC PROCESSING

Closed Die Molding Similar to Die Casting. In a general way, the forming of sheets of plastic may be compared to the pressworking of metals: many of the techniques are similar. Most of the casting methods used with plastics are similar to permanent mold casting of metals. The most important area of plastic processing is matched die molding. In this area, *compression molding* and *cold molding* are like forging and powder metallurgy in that the material is introduced into an open die, and the forming pressure is applied by the closing of the dies. *Transfer molding* is essentially cold chamber die casting, and *injection molding* is quite like hot chamber die casting. In fact, the equipment used for these processes is usually similar in appearance. Extrusion of plastics is directly comparable to the extrusion of metals.

Plastic Type Limits Processing. Many of the procedures have been developed because of the nature of the plastic groups, particularly because of the difference between thermosetting and thermoplastic materials. While the initial treatment of these two types is similar, and both soften during initial heating, this ductile stage of thermosetting plastics is of limited duration, and the setting reaction proceeds with time, particularly at elevated temperature. Thermoplastic materials, however, may be held in the softened condition for prolonged periods of time with little or no chemical change.

COMPRESSION MOLDING

Mold Closing Provides Pressure. The oldest and simplest of plastic molding processes is compression molding, shown in Figure 15-1. Material in powder, granule, pill, or preformed shape is first introduced into the mold, followed by the application of pressure and heat. With thermosetting plastics, for which the process is normally used, the first effect of the heat is to soften the material to a thermoplastic stage

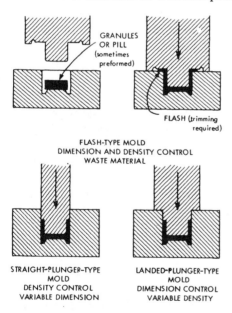

Figure 15-1
Compression mold types

in which the particles coalesce and flow under pressure to fill the mold cavity. With prolonged application of heat, the thermosetting reaction takes place, and the material becomes permanently rigid. The mold may be opened while still hot and the finished part removed, although partial cooling is sometimes beneficial to the dimensional stability of the product. The setting time varies from a few seconds to several minutes, depending on material, temperature, heating method, and section thicknesses. It is possible to compression mold thermoplastics; but, after the pressure and heating portion of the cycle, the mold must be cooled before removal of the part.

Advantages and Limitations of Compression Molding. Compared with other molding techniques, a number of advantages and limitations are associated with compression molding. Size restrictions are relatively few, and the largest molded articles are generally made by this method. There is no waste material and little erosion of the dies because the material does not flow under high pressure from outside the mold. Because of the short, multidirectional flow of material within the mold, distortions and internal stresses within the mold may be minimized. On the other hand, undercuts and small holes are not practical, and the nature of the process requires that the shape of the article be such that the two halves of the mold can fit telescopically together to insure filling. The high pressures required, together with the low viscosity of most thermosetting materials in the plastic state, result in filling clearances between mold parts even when they are on the order of 0.025 millimeter (0.001 inch). Thus, not only will removal of flash from the part be required but also cleaning of the mold parts between successive cycles will frequently be necessary.

CLOSED DIE MOLDING

By far, the most important molding processes used are those that introduce the plastic into closed dies by some external pressure system. The principal difference between these methods and the die casting used in the foundry is the softened plastic condition of the material rather than the liquid state of the metals. Because of the similarities, the terminology is mostly the same as that used in the foundry.

Transfer Molding — Thermosetting Plastics. The variations are due principally to the differences between thermoplastics and thermosetting materials. Transfer molding, used with the latter and shown in Figure 15-2, is like cold chamber die casting in all important respects. A predetermined quantity of molding compound, always including some excess, is introduced into the transfer chamber. This material is usually preformed and may be preheated. Sufficient heat is supplied to the material in the transfer chamber to bring the plastic to the softened state. Pressure is applied to force, or "transfer," the charge to the die cavity. Additional heat is supplied to the die for the thermosetting reaction. The excess material in the transfer chamber and the sprue and runner system also set, resulting in a *cull* that must be removed at the completion of the cycle. This cull is scrap because the thermosetting reaction may not be reversed.

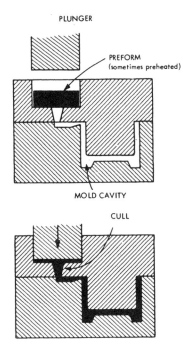

Figure 15-2
Transfer molding

Injection Molding — Thermoplastic Materials. For thermoplastic materials, the transfer process is simplified because of the nature of the material. The term *injection molding* is used to describe the process. Prolonged heating is not necessary or desirable, and the material may be forced into a cool die where the material becomes rigid as a result of cooling rather than chemical change. As indicated in Figure 15-3, a measured charge of raw material is introduced when the plunger is withdrawn, and, on the working stroke of the machine, the material is forced around the spreader where heat is supplied. Material for four to eight working strokes, or *shots*, is normally kept in the heating chamber. Temperatures are controlled so that the sprue separates at the nozzle when the parts are removed, with the material in the nozzle remaining heated sufficiently to be injected on the next cycle without the cull losses normally expected in transfer molding of thermosetting plastics.

Some injection molding of thermosetting materials is done, but precise temperature and time controls are necessary to prevent premature setting of the material in the injection chamber. When used for these materials, the process is known as *jet, flow,* or *offset* molding.

CASTING

With the exception of acrylic rod and sheet materials, which are cast against glass, and some protective coatings applied by dipping, casting of plastics is primarily a low tooling cost procedure restricted to thermosetting resins and used for low production of jewelry, novelty items, laboratory specimens, and similar parts. Polyesters, epoxies, and phenolics are most frequently used in syrupy or liquid form, with hardening promoted by chemical catalysts or by prolonged heating at low temperatures.

EXTRUSION

Most plastics that are finished as sheets, tubes, rods, filaments, films, and other shapes of uniform cross section are produced by extrusion. With some plastics that have a high degree of crystallinity, higher strengths may be developed by stretch deforming the material after extrusion.

Thin Plastic Films. Two methods are used for producing film. In one, the film is extruded through a slit of appropriate size. In the other, the material is extruded as a tube that is then expanded by air pressure and either slit or passed between heated rollers where it is welded into a single sheet. By the expanded tube method, films of less than 0.025-millimeter thickness are produced in large quantities for food wrapping and other packaging.

REINFORCED PLASTIC MOLDING

One of the fastest growing fields in recent years has been the production of relatively large plastic articles with filler in the form of reinforcing fibers in loose, woven, or sheet form. The principle is old; plywood is an example, although the early adhesives used for plywood were not considered to be plastics, and the wood fibers were not fully saturated with resin as is common with most molding of this type now.

Fibrous Fillers — Thermosetting Resins. Glass fibers and paper are the most common filler materials used. Wood and fabric in various forms also have some applications. At present, the process is limited

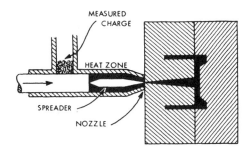

Figure 15-3
Injection molding

to thermosetting materials because of both the nature of the processing used and the higher strengths available. Phenolics, polyesters, melamines, and epoxies predominate.

In nearly all variations of the process, the filler and resin are brought together in the process itself, and the thickness of the molded parts is established more by the placement of the filler material than by mold pressures.

Contact Layup — Filler, Resin. The simplest procedure is *contact layup*, in which successive layers of manually placed filler material are brushed or sprayed with resin as they are applied to the mold, which may have either a concave or a convex shape. The mold may be of almost any material that can be properly shaped, including wood, plaster, concrete, metal, or plastic, and there are almost no size limitations. The resins used may incorporate catalysts that promote setting at room temperatures, or heating may be required. In either case, because no pressure is applied, the ratio of resin to filler must be high to insure complete saturation of the fibers. One of the more interesting applications involves the use of glass filaments, coated with resin, that are wound on mandrels into the shape of spheres or cylinders. With proper winding techniques, the filaments may be orientated to make most efficient use of the longitudinal strength of the fibers; tensile strengths up to 1,000 MPa (150,000 psi) have been reported for structures produced by this method.

Contact Layup Variations. The commonest variations of the contact layup method, *vacuum bag* molding, *expanded bag* molding, and *autoclave* molding, are all methods for developing some pressure on the surface of the molding to permit a lower resin-to-filler ratio. Vacuum bag molding is identical with the contact layup method except that a sheet of vinyl plastic film is placed over the mold after the layers are built up and the mold evacuated to cause atmospheric pressure to be applied. In the expanded bag process, pressures up to 0.35 MPa (50 psi) may be provided by blowing up a bag that conforms to and is held in contact with the molding. The autoclave method is similar to the expanded bag method except that heat and pressure are supplied by steam in a closed chamber.

Compression Process for Sheet Material. In a direct variation of compression molding, matched metal dies are used to form reinforced products. This process is used most for flat sheet manufactured for table and counter tops but is also used for curved shapes, such as chairs, trays, and sinks. For the curved shapes, filler materials are generally preformed before molding. The use of matched metal dies is the only way to produce good finishes on both sides of the finished part, and the high pressures used permit as much as 90% filler and result in higher strengths than would otherwise be possible.

Reinforced Plastics Convenient. The success of "fiberglass" boats, automobile bodies, and similar large shapes attests to the value of reinforced plastics. The simplicity of tooling and equipment required (even for amateur home building projects) makes the contact method ideal for low quantity production and permits rapid design changes when desired. Strength and shock resistance are generally quite high but depend primarily on the type and proportion of filler material.

POSTFORMING

Secondary Operations by Many Methods. Two general classes of operations are performed on plastics after the initial shape has been produced by one of the methods already discussed. Conventional material removal processes, including sawing, shearing, dinking, and blanking, are possible with any plastic but are most frequently used for the preparation of sheet stock prior to a further hot-forming operation. Machining is possible but is generally practical for small quantities only, and other processes are usually cheaper for large quantities. Cutting speeds for thermoplastics must be kept low to prevent heating and softening of the material.

Thermoplastics Often Reheated to Soften. The widest use of postforming operations is made on thermoplastics in sheet form that are heated and made to conform to a single surface mold or pattern by pressure or vacuum. Variations are based primarily on the method of applying pressure and include *draping*, where gravity only is used; *drawing* and *stretch forming*, which are identical to the same operations performed on metal; *blow-dieing*, which is a combined drawing and air-bulging operation; and *vacuum forming*, which is similar to vacuum molding of reinforced plastics except that no external film is used. Some small, relatively flat items, such as brush handles and buttons, are shaped by forging heated sheet stock in closed dies.

DESIGN CONSIDERATIONS

Plastics and Metals Often Competitive. The choice of plastic materials involves the same considerations that apply in choosing metals to fulfill a need. In fact, the two classes of materials are frequently in direct competition with each other. A number of different materials will usually satisfy the functional requirements of a part or product, and the choice depends primarily on the economics of manufacturing for which the material, fabrication, and finishing costs must all be considered. Many plastics require no finishing at all. Often a single plastic molding can replace an assembly of parts made of metal with resulting cost decrease, although the material cost alone may be higher.

Properties of Metals Usually Higher. The stability of properties and the durability of the appearance of plastics are usually poorer than those of metals.

They are generally better for thermosetting materials than for thermoplastics, but the thermosetting plastics are usually slower to process and more expensive. The dimensional stability for plastics ranges from poor to excellent. The low rigidity and thermal conductivity, when compared to metals, may be either advantages or disadvantages, depending on the application.

Plastic strengths are generally lower than metal strengths. Most plastics have tensile strengths below 10,000 psi, but some of the reinforced materials have extremely high strength-to-weight ratios, at higher cost. Many plastic articles compete successfully with metals only through the use of metal inserts for bearings, threads, and fastenings.

Most plastics excel in corrosion resistance to ordinary environments. This is true to the extent that many metals are coated with plastic films for protection.

ADHESIVE BONDING

The elements of an *adhesive bond* are shown in Figure 15-4. An adhesive is most commonly considered to be a material with some "tackiness" or "stickiness," and the animal glues used almost exclusively up to the current century met this requirement. Modern adhesives, however, have wide range in this respect. Contact cements have sufficient tackiness that bonding with considerable strength occurs immediately, under only moderate pressure. Some thermosetting plastic compounds have little or no tackiness as applied and develop strength only after the setting reaction has been promoted by heat, pressure, or chemical reaction with the parts held in place.

Bonding Mechanisms Complex. No clear distinction can be made between the terms *glue, cement,* and *adhesive*. Common to all of them, however, is the property of *adherence* to a surface, and this property is not essentially different from the metallic bond established between metallic surfaces brought into close contact. At least four mechanisms may be responsible for adherence. *Electrostatic* bonds and *covalent* bonds result from the sharing of electrons by different atoms and account for the formation of most common chemical compounds. Even after bonds of these types are established, the positive and negative charges of most atoms are not completely neutralized, and *Van der Waals' forces* provide additional bonding between the atoms. While not strictly an adherence phenomenon, *mechanical interlocking* may take part in the action of some adhesives, although this action appears to be secondary to true adhesion.

Solvents Used with Some Adhesives. As in welding of metals, the proper performance of an adhesive requires that intimate contact be established between the adhesive and the surfaces to be joined. Different means are used to provide closeness. An adhesive can be applied as a solution in a volatile liquid. Evaporation of the solvent is necessary for the adhesive to develop the desired properties, and, as evaporation proceeds, the adhesive proper is drawn to the bare material surfaces. Adhesives of this type are useful for porous materials, such as wood, paper, and fabrics, into which the vapors can penetrate. For nonporous materials, extremely long drying times may be required because the edge of the joint is the only area exposed for evaporation.

Pressure or Heat, or Both, Needed for Some Adhesives. Some relatively new materials are normally solid but become liquid with application of pressure, then resolidify when the pressure is released. Other adhesives are purely thermoplastic in nature, softening or liquifying from heat and hardening on cooling.

Thermosetting Plastic Resins Used for Metals. The most important adhesives for the bonding of metals are thermosetting compounds applied as liquids, pastes, or powders, then polymerized in place through the action of catalysts, heat, or pressure. The materials most used include epoxy, phenolic, polyester, and urea resins.

In addition to the increasing importance of the traditional uses of adhesives in the manufacture of plywood and in the assembly of wood parts, there is considerable growth in the use of adhesives in the bonding of metal structures. These uses are becoming more important as higher strength materials are developed. Adhesives with tensile strengths above 70 MPa (10,000 psi) and shear strengths above 30 MPa (4,000 psi) are available for bonding metals. Many new applications of joining of dissimilar metals, such as rubber to metal, are appearing.

Adhesives Provide Several Advantages. Other advantages may apply to specific cases. Elevated temperatures are not necessary for most adhesives so that distortion associated with welding may be avoided. Thin structures that would be difficult to join by other methods may be used. In most cases, automatic sealing of joints is achieved. This may not be true of mechanically fastened joints. Adhesives may be chosen to provide corrosion resistance or insulation and damping qualities. In many instances, adhesive bonding is used because it does not require expensive equipment and highly trained personnel.

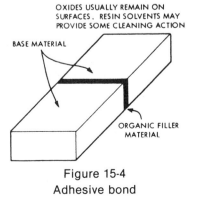

Figure 15-4
Adhesive bond

COMPOSITES

Composites consist of mixtures of two or more materials that maintain their own identities but are attached together in such ways as to reinforce the properties of each by adhesive forces, by their respective positions or frequently by both. Composites may be made up of all metals, combinations of metals and nonmetals or all non-metals. The most usual reason for development of composites has been to produce a lightweight structure with high strength or high stiffness sometimes with the additional feature of withstanding some unusual environmental condition.

LAMINATES

A number of composites are put together in the form of laminates. Most often these are designed in the form of either flat or curved sheets that have very high stength-weight ratios and on the basis of weight may be able to replace steel in many applications. As is true of many new developments in materials and processes, needs of the aerospace industry have initiated the necessary research activities for these materials,.

Aluminum-Boron. The sketch shown in Figure 15-5 displays a typical laminated type composite. Boron, a high melting temperature, hard, strong, non-metallic, in fiber form is enclosed in a unidirectional position in a diffusion bonded aluminum matrix. This core is then sandwiched between two sheets of aluminum alloy. A structure of this kind can be light in weight, but strong and rigid, suitable for many applications in aerospace uses. Problems that could exist are the alignment, spacing, and breakage of the reinforcing fibers and bonding of the interlaminar foils.

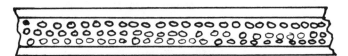

CROSS-SECTION OF COMPOSITE
(50% BORON FILAMENT IN ALUMINUM MATRIX)

Figure 15-5
Aluminum-boron composite

Fiberglass. Glass in fine fiber form is used in a number of ways and for a large number of products as a reinforcing agent for epoxy, polyester, and other thermosetting plastics. Relatively short randomly oriented fibers may be used as a general reinforcing filler. Higher strengths may be obtained by aligning long fibers in the direction the greatest strength is needed. Multidirection alignment may also be used or the fibers may be woven into a cloth before being joined with the plastic. Glass reinforced plastics are used in the construction of boats, automobiles, aerospace vehicle parts, and many other products. To assure that built up laminations are fully bonded in critical applications, careful processing must be maintained and adequate inspections performed.

Graphite. One of the newest developments of strong, lightweight materials is graphite reinforced plastic. Graphite, an amorphous form of carbon, coated with epoxy, is woven into cloth and is molded into sheets, rods, and bars with heat and pressure. The result is a strong, light, flexible product suitable to be laminated into rigid structures such as the aircraft part shown in Figure 15-6 or used in its limber state for such items as fishing rods or golf clubs that are accepted even at high cost.

Figure 15-6
Polyester and fiberglass are used in constructing this boat (*courtesy Fore and Aft*)

Honeycomb. Applications requiring lightweight but strong sheets flat to moderately curved, where allowable space permits use of more than a minor thickness, can often be satisfied by use of honeycomb. Honeycomb derives its strength from a structural design composed of a cellular core of light material encased between two lightweight sheets. Each element is thin, light, and relatively weak, but the combination becomes strong and rigid providing the bonds are sound.

Figure 15-7 shows a simple honeycomb structure of fiberglass and graphite. Manufacturing starts with

Figure 15-7
Honeycomb structure of fiberglass core covered with graphite sheets. All bonded with epoxy.

construction of a sandwich of thin fiberglass sheets and parallel strips of epoxy adhesive. The sandwich is permanently assembled with pressure and heat, then cut to suitable widths and shapes before being pulled open to form the hexagon cells as seen in Figure 15-8.

Figure 15-8
Typical flat sheet honeycomb construction. Hex shaped interior cells formed of thin aluminum strips cemented together with strips of epoxy. After expansion by stretching the assembly, cover plates (one partially left open) are cemented in place to produce a strong, rigid, but light assembly.

Final assembly is accomplished by bonding the thin outside cover sheets to the expanded core using sheets of adhesive and again applying pressure and heat. These bonding operations are frequently accomplished in pressurized ovens called autoclaves.

Careful workmanship with extreme cleanliness are essential for the construction of good honeycomb bonds. Lack of bond or other faults, many of which are illustrated in Figure 15-9, will cause service loads to be transmitted to adjacent cells producing overloads which may result in progressive failure or sudden buckling of the structure. NDT to determine that bond quality is good can be very important. Some, but not all, honeycomb defects are repairable.

The majority of honeycomb used is made with adhesive bonds but some metals, particularly stainless steel, are bonded into this structural form by brazing.

MIXTURES

Some composites are random mixtures of several materials. The properties of such composites may be varied widely by varying the ratios and kinds of constituents used.

Ceramics. Ceramics are produced in a wide variety of types. The majority of ceramics are constructed from clays (compounds of silica and alumina) mixed with water, shaped to proper form, then fired at a fusing temperature in a kiln. Products range from fine china to tile and brick. Ceramics are poor conductors of both heat and electricity. Those that are used in electrical application may require NDT to find cracks and crazes

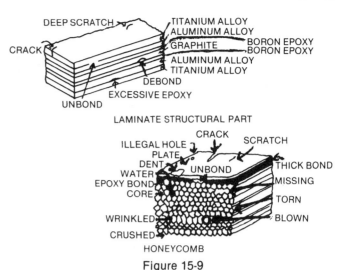

Figure 15-9
Several types of discontinuities in composites

which could hold foreign material and moisture to destroy the insulating property.

Another kind of ceramic, such as used for cutting tool inserts, is made of almost pure alumina (aluminum oxide) assembled from fine particles to a hard rigid block by powder metallurgy methods.

Concrete. A mixture of gravel, sand, and portland cement when combined with enough water to form a thick paste will harden with passage of time into concrete. Concrete is normally used to support compressive loads, but since almost any application, such as the bridge columns and beams of Figure 15-10, is subject to some bending loads (compression and tension), steel reinforcing wires, rods, or structural shapes are nearly always inserted in the material when it is cast.

Portland cement is about 80% carbonate of lime and 20% clay. Additives of various kinds may be added during cement manufacturing to develop special properties. The strength of concrete increases with time. Solidity may occur from a few hours to a few days, but what is defined as 100% strength requires 28 days for standard concrete. Actually strength continues to increase and after one year may reach 150% or more. Most tests performed on concrete are destructive so it is important that proper procedures be used during mixing and pouring.

Rubber. Natural rubber, most of which is imported, is made by coagulating the sap from a rubber tree. Most is vulcanized by combining it with sulphur for stabilization and adding other materials to accentuate certain properties. Most rubber is now classed with similar kinds of properties. The synthetic elastomers are in reality types of plastics and may be developed to be soft and elastic through the entire range to hard and brittle.

The elastomers are used for elastic properties, resistance to many chemicals, resistance to abrasive wear, resistance to slippage, electrical resistance, and many other needs.

Figure 15-10
Bridge structures often appear to be all concrete. Both the columns and beams are internally reinforced to carry any tension laods—usually the tension component of a bending load.

METAL REMOVAL PROCESSES

Discussions of the processes to be covered here are often titled *nontraditional* or *nonconventional* machining. They are certainly nontraditional because they have all been developed since about 1950. Except for the introduction of new tool materials, more sophisticated design, and more highly powered machines, traditional machining has undergone no fundamental changes in the last century. The new processes likewise are nonconventional when compared to conventional machining for they do not necessarily use a high strength tool to cause material failure by applying heavy localized loads to the workpiece.

Most Economically Feasible Only for Special Needs. None of these new methods can currently compete economically with conventional machining for shaping low and moderate strength materials when the surface to be machined is readily accessible and is composed of planes, cylinders, cones, or other simple geometric shapes. However, it is only under special circumstances that materials with hardnesses above about 50 Rockwell C can be machined with single-point cutting tools, and even then tool life is likely to be quite short. In addition, while few shapes are absolutely impossible to machine, many are especially difficult and particularly uneconomical in small quantities. It is toward solving these two problems, high material properties and difficult shapes, that most of these new processes are directed. As with some of the newer low tooling cost pressworking processes, the aerospace industries have been the largest users of these new processes.

Sometimes Referred to as Chipless Machining. These processes are called machining for several reasons. They all remove material, most of them slowly and in small amounts, although not necessarily in chip form. Most of the machines used still have the appearance and general design features of conventional machine tools because they must still provide for the proper positioning of a tool relative to the work and must still provide a geometrically controlled interference path between the tool and the work. The biggest difference occurs in the mechanisms used to produce material failure. With few exceptions, it is a chemical or a thermal, rather than a mechanical, failure.

ELECTRICAL DISCHARGE MACHINING

Old Concept — New Development. The oldest, most successful, versatile, and widely used of the new removal processes is electrical discharge machining, often abbreviated EDM. As early as 1762, it was shown that metals were eroded by spark discharges. Electric arcs have been used to some extent for cutting operations in connection with welding for some time. Practical application to the controlled shaping of metals is much more recent, however, although patents were applied for in the 1930s. The process in its current form dates from about 1950 in this country and a few years earlier in Russia.

High Electrical Voltage Creates Ionized Current Path. The EDM is based on the fact that if an electrical potential exists between two conductive surfaces and the surfaces are brought toward each other, a discharge will occur when the gap is small enough that the potential can cause a breakdown in the medium between the two surfaces. The temperature developed in the gap at the point of discharge will be sufficient to ionize common liquids or gases so that they become highly conductive. It is this ionized column that in the welding process permits a welding arc to be maintained at considerable length, even over short periods of zero voltage when alternating current is used. The condition of maintained ionization is desirable for welding but cannot be tolerated for controlled shaping, as the discharge would tend to remain at one place so long as a low conductive path were present.

Intermittent Direct Current Required. For EDM, the electrodes are separated by a dielectric hydrocarbon oil. The elements of the electrical circuitry are shown in Figure 15-11. A capacitor across the electrodes is charged by a direct-current power supply. With the electrodes separated by about 0.025 millimeter (0.001 inch), a discharge will occur when the voltage reaches 25 to 100 volts, depending on the exact nature of the dielectric and the materials of the electrodes. The essential element of the process is the fact that the discharge will occur at the point where the electrodes are closest together. Whether the discharge should be defined as an arc or a spark is a matter of some debate, but the fact remains that small amounts of material are removed from both electrodes, probably largely as the result of surface vaporization caused by the high temperature developed locally. As soon as the capacitor is discharged, the oil extinguishes the arc (deionizes the path), and the capacitor is then recharged.

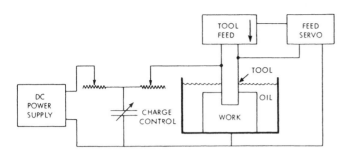

Figure 15-11
Electrical discharge machining

Servomechanism Advances Tool. Subsequent discharges will occur at other points that are then closer together. As material is removed from the electrodes, the distance between them becomes greater, and the voltage required to initiate a discharge rises. This rise in voltage can be used to actuate a servocontrol that feeds the electrodes together to maintain a constant discharge voltage, or stated another way, to maintain a constant distance between the electrodes. The amount of material removed by each discharge will be determined primarily by the amount of energy released from the capacitor. The rate of material removal will be determined by the individual quantity and the cyclic frequency. The frequency of discharges on most machines ranges between 20,000 and 300,000 cycles per second.

Both Workpiece and Tool Are Eroded. The applications for the process depend on the fact that one of the electrodes can be a workpiece, the other a tool that produces a shaped hole, cavity, or external surface in the work. The relative rate of material removal on the work and tool will depend on their melting points, latent heats of evaporation, thermal conductivities, and other factors. Ideally, the material of the tool would be eroded very slowly or not at all. In practice, wear ratios range from as low as 0.05 (twenty times as much workpiece material removed as tool material) when cutting a steel workpiece with a silver tungsten alloy tool, to 2.0 or more when cutting cemented carbides. Because of its low cost and ease of shaping, brass is a more common tool material, although wear ratios are much higher. Graphite provides very favorable wear ratios when used for cutting steel.

Useful for Special Shapes and Hard Materials. The process offers two principal advantages when compared to more traditional methods of machining: some shapes are more easily produced, and workpiece hardness offers no problems (Figure 15-12). The EDM may be used for producing almost any shape if the proper electrode can be made. Noncircular through

Figure 15-12
Electric discharge machine. High frequency electric sparks erode material to the tool shape. The machine can cut hard materials and produce complex shapes but is slow regarding metal removal rates.

holes that would otherwise require a broach or very time-consuming handwork are often made by first removing as much material as possible with a circular drilling operation, then finishing by EDM. The advantage comes in making the electrode because the conventional machining can be done to an external shape. A square or splined electrode, for example, is more easily machined than a square or splined hole if a broach is not available. Electric discharge machining is sometimes a simple and convenient way to fabricate defects in a standard or test specimen for nondestructive testing.

Multiple Duplicated Electrodes Often Needed. If the hole goes through the workpiece, electrode wear creates few problems. The electrode is simply made with additional length that is fed through the work material to compensate for the wear. For a blind hole with straight sides, the electrode would also be made with additional length but would be removed periodically to have its forward end refaced. If the cavity is to have a three-dimensional contour, the problem is more severe. The number of electrodes required would depend on the materials used and on the geometrical precision required. As many as ten electrodes are often used.

All Electrically Conductive Materials Workable. Aside from its ability to cut complex two- or three-dimensional contoured shapes, EDM has the ability to shape any material that has a reasonable amount of electrical conductivity. Hardened steels and cemented carbides present problems no greater than soft ductile materials that could easily be cut by machining. Materials are as easy to shape in a hardened state as they are in an annealed condition.

Slow Removal Process. The process has one drawback in addition to relatively high equipment cost and the problem of electrode wear discussed previously. An inverse relationship exists between the quality of the surface finish produced and the cutting rate. Surface finishes as good as 10 microinches are obtainable, but only with metal removal rates on the order of 0.005 cubic centimeter (0.0003 cubic inch) per minute. Maximum metal removal rates at present are about 5 cubic centimeters (0.3 cubic inch) per minute, but when this rate is achieved, surface finish quality measures about 500 microinches.

ELECTROCHEMICAL MACHINING

Electrochemical machining (ECM) is somewhat newer than EDM but has grown rapidly in the past few years. It offers great potential for the future, particularly because of the greater metal removal rates possible than with EDM.

A Special Reverse Plating System. In this process, as in EDM, both the tool and the workpiece must be conductive, or at least the workpiece must be conductive and the tool must have a conductive coating. With a suitable electrolyte between them, the tool and workpiece form opposite electrodes of an electrolytic cell. The workpiece is connected to the positive terminal of a direct-current supply and the tool to the negative terminal. The electrical circuit is identical to that used in metal plating where metal is removed from the anode and deposited on the cathode.

There are two major differences. Different electrolytes are used so that the material removed from the anode forms insoluble oxides or hydroxides. In electroplating, the unagitated electrolyte permits metal ions to leave the anode only as fast as they can diffuse into the electrolyte. The low rate of diffusion restricts the maximum current flow that can be efficiently used. In ECM the electrolyte is made to flow rapidly between the tool and the work by pressures up to 4 MPa (600 psi). Currents up to 10,000 amperes are used on an area 30 square centimeters (5 square inches) with a resulting metal removal rate of about 16 cubic centimeters (1 cubic inch) per minute. With adequate power supplies, there appears to be no reason that the metal removal rate could not be even greater.

Work Energy Efficiency Low. The ECM is used for many of the same jobs that could be done by EDM, including the making of irregularly shaped holes, forming shaped cavities, and machining very hard or abrasive materials. Figure 15-13 gives an outline of the process. Compared to EDM, tolerances must be greater, particularly in cavity shaping, and tool design is more critical to obtain proper flow of the electrolyte between the tool and the work. In addition, as much as 160 horsepower per cubic inch per minute of metal removal is required. This is about four times that required by EDM, and more than one hundred times that needed by most conventional machining. On the other hand, tools do not wear, and the metal removal rate is much greater than with EDM.

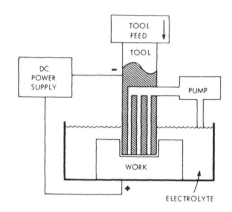

Figure 15-13
Electrochemical machining

CHEMICAL MILLING

This is a process for shaping metals by chemical dissolution without electrical action. The name apparently originated from early applications where the process was used in aircraft manufacture as an adjunct to milling. It was originally used primarily to remove metal for weight reduction in areas of the workpiece that were not accessible to milling cutters and where work contours made following the surface with a cutter virtually impossible.

A Fully Chemical Process. The procedure is relatively simple. The areas of the part where material is not to be removed are first masked with an oxidation-resistant coating. The masking may be done by first coating the workpiece entirely and then removing the masking material from the desired areas by hand. When production quantities warrant, silk screening may be used to apply the maskant only where needed. The part is then immersed in a suitable etchant, which is usually a strong acid or alkali. After the material has been etched to the required depth, the work is removed and rinsed and the maskant removed.

Deep Straight Cuts Impossible. One of the most widely used applications at present is in the manufacture of printed circuit boards for electronic assemblies. The process is also competitive with conventional press blanking for short runs, especially in thin material. One of the principal drawbacks is the undercutting that occurs along the edges of the mask.

Depth control is reasonably good, but straight vertical sides or sharp corners cannot be achieved in the cavity produced.

Variations in circulation of the etchant, variations of temperature, or differences in the material being worked upon may cause variable rates of chemical action. NDT by ultrasonic tests may therefore be necessary on critical parts to check possible thickness variations.

ULTRASONIC MACHINING

A Mechanical Forming Process. The term *ultrasonic machining* is used to denote an abrasive machining process used for cutting hard materials by projecting tiny abrasive particles at the work surface at high velocities. Figure 15-14 shows the details of the process. The abrasive is carried in a liquid flowing between the shaped tool and the workpiece. The tool is made to oscillate along its axis at a frequency of about 20,000 hertz.

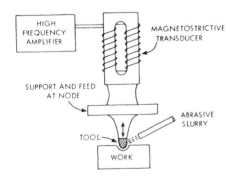

Figure 15-14
Ultrasonic machining

Transducer Motion Amplified by Horn. The heart of the equipment is the *transducer* that converts the high frequency electrical power to mechanical energy. Most transducers are made with nickel laminations that are placed in an oscillating magnetic field. Nickel has the property of *magnetostriction* and undergoes a change in length when placed in a magnetic field. The amplitude of vibration of the nickel is insufficient for practical use and must be amplified by attaching a suitable *horn* to one end. The tool is then brazed, soldered, or mechanically fastened to the end of the horn. The entire assembly must be mechanically tuned to resonate at the frequency produced by the electronic amplifier. When so tuned, the amplitude of the tool motion is from 0.05 to 0.1 millimeter (0.002 to 0.004 inch).

Produces Good Finishes. The tool itself is most often made of soft steel and is given the negative shape of the cavity to be produced, as in EDM or ECM. The most common abrasive used is boron carbide in grit size ranging from 240 (coarse) to 800 (fine). The cutting rate and finish produced both depend on the size of the abrasive. With 800-grit abrasive, finishes as fine as 10 microinches may be attained. Tolerances as close as 0.01 millimeter (0.0005 inch) are possible on size and contour with fine abrasives.

Best for Hard, Brittle Materials. Unlike conventional machining, which works only with material below a certain hardness, and EDM or ECM, which work with any conductive material, ultrasonic machining is best suited to materials that are both hard and brittle. However, the work material need not be a metal or otherwise conductive. The process has been used for engraving, slicing, drilling, and cavity sinking on hardened steel, gem stones, cemented carbides, ferrites, aluminum oxide, glass, and other ceramics.

Not Competitive with Usable Conventional Methods. Metal removal rate is presently the principal drawback, being only about 0.3 cubic millimeter (0.02 cubic inch) per minute. It could possibly be increased considerably with better transducers, but the process is likely to remain in the special-purpose category.

OTHER POSSIBLE MATERIAL REMOVAL METHODS

EDM, ECM, chemical milling, and ultrasonic machining are currently commercially used processes for which equipment is available. Much development work still remains to be done on all these processes, but their current value is sufficient to warrant their existence. Other potential removal processes are now purely in the development stage but may offer competition in the future.

The *laser* (light amplification by stimulated emission of radiation) was invented about 1960. It quickly received much attention and publicity, hailed as the greatest invention of the century. True, it found a number of uses in measuring, in holography, and as a signal carrier but, due mainly to limited capacity and high cost, was until recently of little value for either machining or welding on a commercial scale.

Development since 1970 has increased the power capability and reduced the cost to make it more competitive with conventional equipment. Some hole making and cutting, as well as other type of operations, are being performed industrially. It is predicted that industrial laser use will grow at a rate of about $20 million per year. The uses continue to be most with materials difficult to manufacture with the more common methods.

The plasma arc was discussed in Chapter 10 together with electron beams as a heating source for welding. Plasma arcs are also capable of sufficiently localized energy inputs that surface material may be melted and vaporized with relatively small heating of the adjacent material. The arcs are being used for some straight-line cutting operations, where control is simple and tolerance requirements are not too high. Some experimental work has been done in lathe turning, using a plasma arc as a cutting tool.

DEPOSITION PROCESSES

Of the more traditional processes, both welding processes using filler material and casting involve the deposition of molten material. The material is forced to conform to the desired shape by pressure provided by gravity, external pressure, or surface tension. One newer process of the same general type deposits material in controlled small amounts and permits a shape to be built up. Another new process is a variation of powder metallurgy that allows more complex shapes to be produced and does not require conventional dies.

ELECTROFORMING

This process may be described as the reverse of ECM. When a direct current is passed between two electrodes immersed in the proper electrolyte, material is removed from the anode and deposited on the cathode. This action is the basis of electroplating, which will be discussed later. For electroforming, however, coatings of much greater thickness, up to 10 millimeters (3/8 inch), are built up.

Electrical Conductance of Pattern Essential. For the production of an electroformed part, a master, or pattern, must first be produced with external shape and dimensions corresponding to the interior shape desired in the work. The pattern must have a conducting surface. If made of a nonconducting material, it must be coated with a conducting film of metal or graphite. The pattern is then placed in the electrolyte and the metal deposited to the required thickness. For certain shapes, the part may be stripped from the pattern and the pattern reused. Other shapes may require that the pattern be removed chemically or, if made of a low melting point material, by melting.

Complex Shapes and Miniature Parts Possible. The process has a number of advantages. It is possible to produce complex internal contours with close dimensional control and surface finishes as good as 8 microinches. Because of these properties, electroforming is used in making high frequency wave guides and venturis for nozzles and flow measurement. Parts may be made much thinner than by most conventional processes. It is possible to deposit most metals by the process. Parts with different metals on the interior and exterior surfaces may also be produced.

On the other hand, wall thickness is difficult to keep uniform so that exterior shapes and dimensions may not be controlled accurately. As with chemical milling, critical parts may call for thickness checks by ultrasonic tests. The rate of production is normally quite slow and the cost is high.

GROSS SEPARATION PROCESSES

The following processes are "miscellaneous" only because they do not fit well in any of the established categories of casting, welding, deformation, or machining discussed previously. For many applications they are in direct competition with sawing and shearing for both straight line and contour cutting.

TORCH CUTTING

This separation process depends on keeping the material being cut above its kindling temperature (800°C or 1500°F for pure iron) and supplying a stream of oxygen to promote fast oxidation. High temperature in the cutting zone is aided by the exothermic reaction of burning material.

Process Limited Mostly to Steels. Conditions for cutting are easily obtained with pure iron and low alloy steels but are different with many other metals. Reduced exothermic reaction and/or increased thermal conductivity reduce the practicality of using the process with cast iron, high alloy steels including stainless and most nonferrous alloys.

Easily Mechanized. Figure 15-15 shows a mechanized setup for cutting a straight line cut in steel plate. Oxyacetylene flames are used to bring the steel to kindling temperature, then pure oxygen is supplied through a central orifice in the torch tip to burn a slot through the steel as the carriage moves along as its guide. The torch path may also be established by numerical control or may be guided by a line reader following the lines on a part drawing.

Figure 15-15
Oxyacetylene cutting

The process is very versatile, may be equipped with multiple torches for higher production, and produces accuracies similar to those obtained by sawing. This sheet may be cut singly or stacked. Steel over 5 feet in thickness has been cut by this process and *scarfing*, removal of defects in large casting and forgings, is commonly practiced by use of flame cutting.

Arc Cutting Possible. Use of a steel wire electrode fed at high speed with gas shielding and very high currents can also be used for cutting. Thickness of cut is much more limited than with the torch method, but materials difficult to cut with flame can be parted with the arc.

FRICTION SAWING

Friction sawing has limited but important use. This process also is used most for cutting steel.

High Speed Rubbing Creates Heat. Localized heat is created in the workpiece by contact with the edge of a fast moving blade or disc. Edge speeds are in the range of 3000 to 7500 meters per minute (15000-25000 fpm). The tool may be smooth edged but usually has notches or teeth that help remove softened metal from the kerf.

The process is used mostly for cutoff work on bars and structural shapes in steel mill and warehouse operations. It may also be useful occasionally for cutting steel that is too hard to be cut by conventional means.

Surface Finishing 16

Products that have been completed to their proper shape and size frequently require some type of surface finishing to enable them to satisfactorily fulfill their function. In some cases, it is necessary to improve the physical properties of the surface material for resistance to penetration or abrasion. In many manufacturing processes, the product surface is left with dirt, chips, grease, or other harmful material upon it. Assemblies that are made of different materials, or from the same materials processed in different manners, may require some special surface treatment to provide uniformity of appearance.

Surface finishing may sometimes become an intermediate step in processing. For instance, cleaning and polishing are usually essential before any kind of plating process. Some of the cleaning procedures are also used for improving surface smoothness on mating parts and for removing burrs and sharp corners, which might be harmful in later use. Another important need for surface finishing is for corrosion protection in a variety of environments. The type of protection provided will depend largely upon the anticipated exposure, with due consideration to the material being protected and the economic factors involved.

Satisfying the above objectives necessitates the use of many surface-finishing methods that involve chemical change of the surface, mechanical work affecting surface properties, cleaning by a variety of methods, and the application of protective coatings, organic and metallic.

CASEHARDENING OF STEELS

Casehardening Results in a Hard, Shell-like Surface. Some product applications require surface properties of hardness and strength to resist penetration under high pressure and to provide maximum wear qualities. Where through hardness and the maximum strength associated with it are not necessary, it may be more economical to gain the needed surface qualities by a *casehardening* process. Casehardening involves a change of surface properties to produce a hard, wear-resistant shell about a tough, fracture-resistant core. This is usually accomplished by a change of surface material chemistry. With some materials, a similar condition can be produced by a phase change of material already present.

Multiple Benefits from Casehardening. Casehardening may be more satisfactory than through hardening in those cases where a low cost, low carbon steel with a hard shell may be used instead of a higher cost, high carbon or alloy steel needed for through hardening. The process is much less likely to cause warping or cracking and the product, because of its soft, ductile core, is less subject to brittle failure than a through-hardened product. Casehardening is often suitable for heavy sections that would require very special high alloy steels for through hardening to be effective.

Case depth measurement is sometimes checked by destructive methods: cutting the object, etching the cut surface, and checking the cut depth with a measuring microscope. A faster and more usable method when knowledge is needed directly for service parts is by use of eddy current tests.

CARBURIZING

Casehardening of steel may be accomplished by a number of methods. Choice between them is dependent on the material to be treated, the application, and the desired properties. One of the more common methods is carburizing, which implies an increase or addition of carbon which is actually the basis of the process.

Performed on Low Carbon Steels. Carburizing is usually performed on a low alloy or plain low carbon steel. If an alloy steel is used, it usually contains small quantities of nickel or some other element that acts as a retardant to grain growth during the heating cycle. Low carbon steels are commonly used to minimize the effect of subsequent heat treatments on the core material. It is possible to carburize any steel containing less than the 0.70% to 1.20% carbon that is produced in the surface material.

Carbon Diffusion Is Time-Temperature Dependent. Carbon is caused to diffuse into the steel by heating the material above its critical temperature and holding it in the presence of excess carbon. Temperatures used are usually between 850°C and 930°C, with the choice most dependent on the desired rate of penetration, the desired surface carbon content, and the permissible grain growth in the material. Penetration is dependent upon both the temperature and time, with variation of case depth from 0.25 to 1.0 millimeter (0.010 to 0.040 inch) possible in the first 2 hours by varying the temperature between the two extremes. The rate of penetration slows down as the depth increases, as shown in Figure 16-1, so that for large depths, relatively long periods of time are necessary.

Carbon May Be Supplied from a Gas, Liquid, or Solid Environment. The excess carbon for diffusion is supplied from a carbon-rich environment in solid, liquid, or gas form. Parts to be carburized may be packed in carbon or other carbonaceous material in

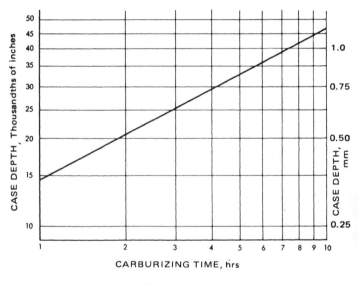

Figure 16-1
Typical carburizing case depth-time relationship

boxes that are sealed to exclude air and then heated in a furnace for the required length of time, in a process sometimes referred to as *pack hardening*. The liquid method makes use of molten sodium cyanide in which the parts are suspended to take on carbon. The cyanide method is usually limited to shallow case depths of about 0.25 millimeter (0.010 inch) maximum. The third method — often the most simple for production operations requiring heavy case depth — supplies gaseous hydrocarbons from an unburned gas or oil fuel source to the furnace retort in which the product is heated. The product is usually suspended on wires or rolled about in order that all surfaces will be exposed uniformly.

Grain-Size Control Necessary for Best Properties. Carburizing steels containing grain-growth inhibitors may be quenched directly from the carburizing furnace to harden the outside shell, but plain carbon steels must be cooled and reheated through the critical temperature range to reduce grain size. Even the alloy steels will have better properties if treated in this manner. Quenching from above the critical temperature will produce a hard martensitic structure in the high carbon surface material but will have little or no effect on the low carbon core. As in the case of most through-hardened steels, tempering is usually required to toughen the outside shell. The complete cycle for casehardening by carburizing is illustrated in Figure 16-2.

FLAME HARDENING

Surface Must Be Heated above Transformation Temperature. Another casehardening process that does not require a change of composition in the surface material is flame hardening. This method can

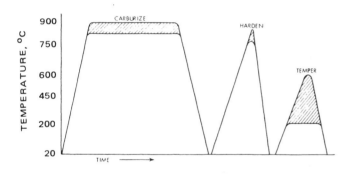

Figure 16-2
Heating cycle for casehardening by carburizing

be used only on steels that contain sufficient carbon to be hardenable by standard heat-treating procedures. The case is produced by selectively heating part or all of the surface with special high capacity gas burners or oxyacetylene torches at a rate sufficiently high that only a small depth from the surface goes above the critical temperature. Following immediately behind the torch is a water quenching head that floods the surface to reduce the temperature fast enough to produce a martensitic structure. As in the case of carburizing, the surface may be then reheated to temper it for toughness improvement. The depth of hardness is controlled by the temperature to which the metal is raised, by the rate of heating, and by the time that passes before quenching.

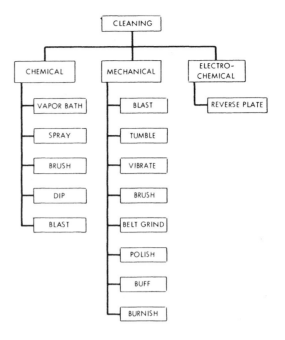

CLEANING

Few, if any, shaping and sizing processes produce products that are usable without some type of cleaning unless special precautions are taken. Hot working, heat treating, and welding cause oxidation and scale formation from high temperature in the presence of oxygen. For the same reason, castings usually are coated with scale or oxides. If made in sand molds, they may have sand grains fused or adhering to the surface. Residue from coolants, lubricants, and other processing materials is common on many manufactured parts. In addition to greasy films from processing, protective coatings of greases, oils, or waxes are frequently used intentionally to prevent rust or corrosion on parts that are stored for some period of time before being put to use. Even if parts are clean at the completion of manufacturing, they seldom remain that way for long. After only short storage periods, corrosion and dust from atmospheric exposure necessitate cleaning for best condition or to permit further processing.

When using NDT such as penetrant testing and ultrasonic testing, good precleaning may be necessary to get accurate results and postcleaning is often needed to leave the surface in suitable condition. In some applications, such as on stainless steels and nickel based alloys, ultrasonic coolants and penetrant materials must be made of only certain material so that the NDT materials are not one of the causes of a stress corrosion failure.

Cleaning sometimes has finish improvement associated with it. Some shape-producing methods produce unsatisfactory surface characteristics such as sharp corners, burrs, and tool marks, which may affect the function, handling ease, and appearance of the product. Some cleaning processes at least partially blend together surface irregularities to produce uniform light reflection. Improvement of surface qualities may be accomplished by removal of high spots by cutting or by plastic flow as cleaning is performed.

CHOICE OF CLEANING METHOD

As indicated by the list at the head of this section, many different cleaning methods are available. The one most suitable for any particular situation is dependent upon a number of factors. Cost is, of course, always a strong consideration, but the reason for cleaning is bound to affect the choice. Convenience in handling, improvement in appearance, elimination of foreign material that may affect function, or establishment of a chemically clean surface as an intermediate step in processing might all call for different methods. Consideration must be given to the starting conditions and to the degree of improvement desired or required. Methods suitable for some materials are not at all satisfactory for use on other kinds of material.

Cleaning and Corrosion Protection Sometimes Associated. Some cleaning methods provide multiple benefits. As pointed out previously, cleaning and finish improvement are often combined. Probably of even greater importance is the combination of corrosion protection with finish improvement, although corrosion protection is more often a second step that involves coating an already cleaned surface with some other material.

LIQUID AND VAPOR BATHS

Liquid and Vapor Solvents Common. The most widely used cleaning methods make use of a cleaning medium in liquid or vapor form. These methods depend on a solvent or chemical action between the surface contaminants and the cleaning material. Many cleaning methods and a variety of materials are available for choice, depending on the base material to be cleaned, the contaminant to be removed, the importance and degree of cleanliness, and the quantity to be treated.

Petroleum Solvents Good for Greases and Oils. Among the more common cleaning jobs required is the removal of grease and oil deposited during manufacturing or intentionally coated on the work to provide protection. One of the most efficient ways to remove this material is by use of solvents that dissolve the grease and oil but have no effect on the base metal. Petroleum derivatives such as Stoddard solvent and kerosene are common for this purpose, but as they introduce some danger of fire, chlorinated solvents, such as trichretholene, that are free of this fault are sometimes substituted.

Conditioned Water Usually Inexpensive. One of the most economical cleaning materials is water. However, it is seldom used alone even if the contaminant is fully water soluble because the impurity of the water itself may contaminate the work surface. Depending on its use, water is treated with various acids and alkalies to suit the job being performed.

Proper Pickling Can Selectively Remove Iron Oxides. Water containing sulfuric acid in a concentration from about 10% to 25% and at a temperature of approximately 65°C is commonly used in a process called *pickling* for removal of surface oxides or scale on iron and steel. The work is immersed in the solution contained in large tanks for a predetermined period of time after which it is rinsed to stop the chemical action.

Improper control of the timing, temperature, or concentration in the pickling bath is likely to result in pitting of the surface because of uneven chemical reaction. Most pickling baths are treated with chemical inhibitors that decrease the chemical effect of the acid on the base metal but have little effect on the rate at which the oxides are attacked.

Many Water Additives Are Proprietary Mixtures. Many of the common cleaning liquids are made up of approximately 95% water containing alkaline cleaners such as caustic soda, sodium carbonate, silicates, phosphates, and borates. The proportions are varied for different purposes and are available under different brand names for particular applications.

Application Dependent on Material and Purpose. Liquid cleaners may be applied in a number of ways. Degreasing, particularly on small parts, is frequently done with a vapor bath. This does an excellent job of removing the grease but has the disadvantage of not being able to remove chips and other kinds of dirt that might be present. Vapor degreasing is usually done in a special tank that is heated at the bottom to vaporize the solvent and cooled at the top to condense the solvent. Cold work suspended in the vapor causes condensation of the solvent, which dissolves the grease and drips back into the bottom of the tank. The difference in volatility between the solvent and the greases permits the vapor to remain unchanged and to do a uniform cleaning job.

Mechanical Work Frequently Combined with Chemical Action. Spraying, brushing, and dipping methods are also used with liquid cleaners. In nearly all cases, mechanical work to cause surface film breakdown and particle movement is combined with chemical and solvent action. The mechanical work may be agitation of the product as in dipping, movement of the cleaning agent as in spraying, or use of a third element as in rubbing or brushing. In some applications, sonic or ultrasonic vibrations are applied to either the solution or the workpieces to speed the cleaning action. Chemical activity is increased with higher temperatures and optimum concentration of the cleaning agent, both of which must in some cases be controlled closely for efficient action.

Important That Chemicals Be Removed. Washing and rinsing away of the cleaning liquids is usually necessary to prevent films and spots. Fast drying of water solutions on iron and steel products is sometimes needed to prevent the formation of rust. If the product mass is large enough, heat picked up from the cleaning bath may be sufficient to cause fast drying; otherwise, air blasts or external heat sources may be required.

BLASTING

Blasting Provides Large Mechanical Action. The term *blasting* is used to refer to all of those cleaning methods in which the cleaning medium is accelerated to high velocity and impinged against the surface to be cleaned. The high velocity may be provided by air or water directed through a nozzle or by mechanical means with a revolving slinger. The cleaning agent may be either dry or wet solid media such as sand, abrasive, steel grit, or shot, or may be liquid or vapor solvents combined with abrasive material.

Operator Safety Must Be Considered. The solid media are used for the removal of brittle surface contamination such as the heat-treat scale found on forgings and castings. Steel grit has replaced sand and other refractory-type abrasives to some extent because of the reduced health hazard (silicosis) and a reduced tendency for pulverization. Sand, however, can be used without danger to the operator when parts are small enough to be handled by hand inside a properly designed chamber fitted with a dust collector.

Surface Stressed and Work Hardened. In addition to cleaning, solid particles can improve finish and surface properties of the material on which they are used. Blasting tends to increase the surface area and thus set up compressive stresses that may cause a warping of thin sections, but in other cases, it may be very beneficial by reducing the likelihood of fatigue failure. When used for this latter purpose, the process is more commonly known as *shotpeening*.

Water Slurries. Liquid or vaporized solvents may, by themselves, be blasted against a surface for high speed cleaning of oil and grease films with both chemical and mechanical action. Water containing rust-inhibiting chemicals may carry, in suspension, fine abrasive particles that provide a grinding cutting-type action for finish improvement along with cleaning. The blasting method using this medium is commonly known as *liquid honing*.

ABRASIVE BARREL FINISHING

A Low Cost Cleaning and Finishing Method. When large numbers of small parts that do not need to have sharp detail or accurate dimensions require cleaning, the rotating barrel method may be very economical. Names used are: *Barrel finishing, rolling, tumbling,* and *rattling*. They are all similar but various media may be combined with the work as indicated in Figure 16-3. High polish may be produced by tumbling with pieces of leather to wipe the surfaces smooth as in a strop honing operation. In some cases a number of hours may be required to produce the desired results but since the finishing machines do not have to be tended by operators, the unit cost may be extremely low.

Machines with a vibratory motion and loaded with abrasive media are also used for similar type cleaning and finishing work.

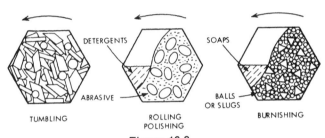

Figure 16-3
Barrel finishing

WIRE BRUSHING

A number of cleaning operations can be quickly and easily performed by use of a high speed rotating wire brush. In addition to cleaning, the contact and rubbing of the wire ends across the work surface produces surface improvement by a burnishing-type action. Sharp edges and burrs can be removed. Scratches, rough spots, and similar mechanical imperfections can be improved primarily by plastic flow, which also tends to work harden the surface material. Most wire brushing is done under manual control, but where the surfaces can be made accessible and the quantity to be treated is sufficiently large for economic feasibility, machines for automatic brushing can be set up.

Common applications of wire brushing are the cleaning of castings, both ferrous and nonferrous; the cleaning of spatter and slag from weldments; and the removal of rust, corrosion, and paint from any object whose base material is strong enough to withstand the brushing. Wire brushing produces a distinctive pattern on the surface and in addition to cleaning, it sometimes is used to produce a decorative surface.

A precaution regarding surface defect detection should be kept in mind. *Any* method of surface cleaning involving abrasion or rubbing may smear the surface material in such a way as to disguise or cover over surface defects and prevent their detection by usual methods. Careful selection of a method may be necessary, or in some cases, such drastic methods as etching may be needed. Machining, inlcuding fine grinding, also has similar effects to a lesser degree but should be remembered when small defects could be serious regarding service life of the part under consideration. Penetrant tests are most severely affected and can be rendered practically useless if defect openings have been smeared.

POLISHING

The term *polishing* may be interpreted to mean any nonprecision procedure providing a glossy surface but is most commonly used to refer to a surface-finishing process using a flexible abrasive wheel. The wheels may be constructed of felt or rubber with an abrasive band, of multiple coated abrasive discs, of leaves of coated abrasive, of felt or fabric to which loose abrasive is added as needed, or of abrasives in a rubber matrix.

Polishing Is a Surface Blending Process. These wheels differ from grinding wheels only by being flexible, which enables them to apply uniform pressure to the work surface and permits them to conform to the surface shape.

Polishing is usually done offhand except when the quantity is large. The process may have several objectives. Interest may be only in finish improvement for appearance. The surface finish may be important as an underlay for plating, which has only limited ability to improve surface quality over that of the surface on which it is placed. Polishing may also be important as a means of improving fatigue resistance for products subject to this kind of failure.

BUFFING

About the only difference between buffing and polishing is that, for buffing, a fine abrasive carried in wax or a similar substance is charged on the surface of a flexible wheel. The obejctives are similar. With finer abrasive, buffing produces higher quality finish and luster but removes only minor amounts of metal. With both polishing and buffing, particularly of the softer metals, plastic flow permits filling of pores, scratches, and other surface flaws to improve both appearance and resistance to corrosion.

ELECTROPOLISHING

If a workpiece is suspended in an electrolyte and connected to the anode in an electrical circuit, it will supply metal to the electrolyte in a reverse plating process. Material will be removed faster from the high spots of the surface than from the depressions and will thereby increase the average smoothness. The cost of the process is prohibitive for very rough surfaces because larger amounts of metal must be removed to improve surface finish than would be necessary for the same degree of improvement by mechanical polishing. Electropolishing is economical only for improving a surface that is already good or for polishing complex and irregular shapes, the surfaces of which are not accessible to mechanical polishing and buffing equipment.

COATINGS

Many products, in particular those exposed to view and those subject to change by the environment with which they are in contact, need some type of coating for improved appearance or for protection from chemical attack. All newly created surfaces are subject to corrosion, although the rate of occurrence varies greatly with the material, the environment, and the conditions. For all practical purposes, some materials are highly corrosion resistant because the products of corrosion resist further corrosion. For example, a newly machined surface on an aluminum alloy will immediately be attacked by oxygen in the air. The initial aluminum oxide coating protects the remaining metal and practically stops corrosion unless an environmental change occurs. Corrosion rates are closely dependent on environment. Rates increase with rise of temperature and greater concentration of the attacking chemical.

Corrosion Deteriorates Appearance and Properties. The need for corrosion protection for maintenance of appearance is obvious. Unless protected, an object made of bright steel will begin to show rust in a few hours of exposure to ordinary atmosphere. In addition to change of appearance, loss of actual material, change of dimensions, and decrease of strength, corrosion may be the cause of eventual loss of service or failure of a product. Material that must carry loads in structural applications, especially when the loads are cyclic in nature, may fail with fatigue if corrosion is allowed to take place. Corrosion occurs more readily in highly stressed material where it attacks grain boundaries in such a way as to form points of stress concentration that may be nuclei for fatigue failure.

Corrosion Reduced by Proper Design. The corrections for corrosion problems include choice of materials that resist attack from the environment to which they are exposed, selection or control of the environment to minimize corrosion effects, and the use of selective corrosion by placing materials with greater susceptibility to corrosion near those to be protected. The latter is illustrated by the use of magnesium rods in hot water tanks. The magnesium is the target for corrosion; as long as it is present, corrosion of the steel walls of the tank is insignificant. Another correction for corrosion, when the others are impractical, is the coating of the surfaces needing protection with a material that excludes the environmental elements that are harmful.

Thickness of coatings may be important for many reasons. If the objective is improvement of appearance, uniformity of coating may be required, or lacking that some minimum value may have to be surpassed to provide the appearance of uniformity. Life of a coating is usually also closely associated with uniformity and depth of coating layer. Many coatings are inherently porous to some degree and resistance to corrosion is likely to require thickness sufficient to resist penetration of liquids and gases. For those reasons manufacturing specifications frequently list minimum thickness for coatings and a NDT measurement is usually the only way to know when that specification is being met. Although other methods are possible, gaging with eddy current methods is common.

Many Coatings Improve Appearance. In addition to stabilizing appearance by resisting corrosion, coatings are often very valuable for providing color control, change in appearance, and variety, which may be important to sales appeal. Some coatings, such as fillers, paint, and others with substantial body, improve surface smoothness by filling pores and cavities. Some coating materials can provide uniform appearance for products made as assemblies of different materials.

Some Coatings Improve Properties. Coatings of various types may be used to change or improve surface properties. Casehardening of steel has been discussed earlier, and although it is a surface property-changing method, in most of its forms, casehardening does not consist of the addition of a coating.

Wear Resistance by Plating. Hardness and wear resistance can, however, be provided on a surface by plating with hard metals. Chromium plating of gages subject to abrasion is frequently used to increase their wear life. Coatings of plastic materials and asphaltic mixtures are sometimes placed on surfaces to provide sound deadening. The additional benefit of protection from corrosion is usually acquired at the same time.

Increase or Decrease of Coefficient of Friction. Friction characteristics of a surface can be varied in either direction by application of a coating. Rubber and some other plastic materials may be applied for increase of friction characteristics. An example would be the special compounds applied to the floorboards or bottom of small boats to decrease the chance of slipping. Other plastic materials, the fluorocarbons being good examples, are applied to surfaces where slipping is required because they provide a very low coefficient of friction.

PREPARATION FOR COATINGS

Adhesion Associated with Cleanliness. The ability of an organic film to adhere to a metal surface (adhesion) is dependent to a large degree on the cleanliness of the metal surface. However, some materials hold together tighter on a surface that has been slightly roughened by some process such as sand blasting, while others may require chemical treatment of the base metal for formation of an oxide or phosphate film for satisfactory adhesion.

Cleaning by one or more of the methods discussed earlier in this chapter is usually essential before any kind of coating should be applied. In practically every case a clean dry surface is necessary for coating adhesion. Whether or not a combination cleaning and smoothing operation should be used depends somewhat on the previous processing as well as on the desired final finish. Some coatings, such as the heavier plastics, can hide large faults and surface imperfections, but others, such as finishing lacquers and metallic platings, improve finish quality to only a very small degree. With the latter, scratches, surface faults, and even tool marks can continue to show on the final surface although the coating tends to blend and soften their appearance.

PAINTS, VARNISHES, AND ENAMELS

Painting is a generic term that has come to mean the application of almost any kind of organic coating by any method.

Paint. As originally defined and as used most at present, paint is a mixture of pigment in a drying oil. Color and opacity are supplied by the pigment. The oil serves as a carrier for the pigment and in addition creates a tough continuous film as it dries.

Varnish Is Normally Clear. *Varnish* is a combination of natural or synthetic resins and drying oil, sometimes containing volatile solvents as well. The material dries by a chemical reaction in the drying oil to a clear or slightly amber-colored film. A solution of resin in a volatile solvent without the drying oil is called *spirit*, or *shellac*, varnish.

Pigment in Varnish Creates Enamel. *Enamel* is a mixture of pigment in varnish. The resins in the varnish cause the material to dry to a smoother, harder, and glossier surface than produced by ordinary paints. Some enamels are made with thermosetting resins that must be baked for complete dryness. These *baking enamels* provide a toughness and durability not usually available with the ordinary paints and enamels.

LACQUERS

Lacquers Easily Removed. The term *lacquer* is used to refer to finishes consisting of thermoplastic materials dissolved in fast drying solvents. One common combination is cellulose nitrate dissolved in butyl acetate. Present-day lacquers are strictly air drying and form films very quickly after being applied, usually by spraying. No chemical change occurs during the hardening of lacquers, consequently, the dry film can be redissolved in the thinner. Cellulose acetate is used in place of cellulose nitrate in some lacquers because it is nonflammable. Vinyls, chlorinated hydrocarbons, acrylics, and other synthetic thermoplastic resins are also used in the manufacture of lacquers.

Common Because of Fast Drying. Clear lacquers are used to some extent as protective films on such materials as polished brass, but the majority are pigmented and used as color coats. The pigmented lacquers are sometimes called *lacquer enamels*. Lacquers are widely used for coating manufactured products because of their ease of application and speed of drying.

ORGANIC COATING APPLICATION

Paint-type materials are applied by dip, brush, and spray.

Minimum Labor Cost by Dipping. Dipping is common for applying protective coatings to forgings and castings to prevent rust during storage and processing and to serve as primers for the final finish. Many other products made in large quantities also are finished by dipping. Dip application is limited to parts that do not have recesses, pockets, or shapes that will hold the liquid paint or prevent its flowing to an even coat.

Brushing Costly. Brush painting is slow and used little in manufacturing work except on large, heavy, or odd-shaped parts that cannot be moved or manipulated in a spray-paint area. Brushing and rolling are commonly used for coating structural surfaces such as walls and ceilings of buildings. Brushing does provide efficient use of coating material, as practically none is wasted, and the mechanical rubbing of a brush or roller provides some cleaning action that may provide better adhesion.

Speed and Quality by Spraying. By far the greatest amount of organic coatings are applied industrially by spraying. This method is used most with lacquers and fast drying enamels. The short drying time causes parts to become dust free very quickly so that they can be moved away from the spray area and advantage can be taken of this fast application method. Spraying is done in booths designed for this purpose where adequate ventilation carries fumes and spray particles away from the operator (Figure 16-4).

Figure 16-4
Paint spraying in a booth where exhaust air draws the waste paint and fumes away from the operator through filters to clean the air

Spray painting of automobile bodies and other large objects that are conveyorized is often done automatically with a number of spray heads, some stationary and some movable, adjusted to spray a uniform layer over the entire object.

In many cases spray application of penetrant materials is the fastest and best way of obtaining uniform coverage. Spraying aids particularly on parts containing recesses and corners difficult to contact with a brush.

Uniform Coating by Electrostatic Spraying. For electrostatic spraying the paint particles are sprayed through a high voltage electrostatic field. Each paint particle takes on an electric charge from the field and is attracted toward the grounded article to be painted. This method provides better efficiency of paint use than ordinary spraying, but even more important, causes the coating to distribute itself more evenly over the entire object. Electrostatic force can also be used to pull off drips or tears that form by gravity along the bottom edges of newly painted objects.

Heat Often Used to Speed Drying. As indicated previously, organic coating is often done in free air. Some solvents and vehicles are so volatile that drying is accomplished almost immediately. Others require several days for drying, and still others require elevated temperatures for necessary polymerization to take place. Heat for drying and speeding chemical reaction may be provided by various types of ovens. Some ovens are batch types in which racks of parts are placed for specific periods of time. Others are continuous types built over conveyor systems that regulate the time of exposure by the length of oven and the speed of conveyor operation.

VITREOUS ENAMELS

Porcelain Consists of Fused Glass. Vitreous, or *porcelain*, enamel is actually a thin layer of glass fused onto the surface of a metal, usually steel or iron. Shattered glass, ball milled in a fine particle size, is called *frit*. Frit is mixed with clay, water, and metal oxides, which produce the desired color, to form a thin slurry called *slip*. This is applied to the prepared metal surface by dipping or spraying and, after drying, is fired at approximately 800°C to fuse the material to the metal surface. For high quality coating, more than one layer is applied to guard against pinhole porosity.

Excellent Corrosion Protection. Glass applied in this way has high strength and is usually flexible enough to withstand bending of the steel within the elastic limits of the base metal. The coatings have excellent resistance to atmospheric corrosion and to most acids. Vitreous enamels can be made suitable for use over a wide range of temperatures. Some special types have been used for corrosion protection on exhaust stacks for aircraft engines. Considering their high quality protection, vitreous enamels are relatively inexpensive and find many uses.

Ceramic Coatings for Special Protection. The advent of rockets and missiles has introduced an entirely new field in which high temperature corrosion protection is essential. Porcelain enamel has been satisfactory in some of these applications, but ceramic coatings with better refractory characteristics are more commonly used. Some are applied in the same way as porcelain enamel. Others are fused to the metal surfaces with the intense heat of a plasma jet.

Porosity of porcelain or ceramic coatings can be checked with penetrants and coating thickness determined by use of eddy current methods.

METALLIZING

Metal spraying, or *metallizing*, is a process in which metal wire or powder is fed into an oxyacetylene heating flame and then, after melting, is carried by high velocity air to be impinged against the work surface. The small droplets adhere to the surface and bond together to build up a coating.

Bond Mostly Mechanical. The nature of the bond is dependent largely on the materials. The droplets are relatively cool when they make contact and can in fact be sprayed on wood, leather, and other flammable materials. Little, if any, liquid flow aids the bonding action. If, however, sufficient affinity exists between the metals, a type of weld involving atomic bonds may be established. The bond is largely mechanical in most cases, and metal spraying is usually done on surfaces that have been intentionally roughened to aid the mechanical attachment.

Anodic Materials Cause Selective Corrosion. Zinc, aluminum, and cadmium, which are anodic to steel and therefore provide preferential corrosion protection, are usually sprayed in thin layers, averaging about 0.25 millimeter (0.010 inch) in thickness, as protective coatings. Because sprayed coatings tend to be porous, coatings of two or more times this thickness are used for cathodic materials such as tin, lead, and nickel. The cathodic materials protect only by isolating the base material from its environment.

Buildup by Metal Spraying. Another important application for metal spraying is in salvage operations for which a wide variety of metals and alloys may be used. Surfaces, usually after first being roughened, are built up to oversized dimensions with metal spray. The excess material is then machined away to the desired dimension. Expensive parts with worn bearing surfaces or new parts that have been machined undersized can sometimes be salvaged by this relatively cheap procedure.

VACUUM METALLIZING

Some metals can be deposited in very thin films, usually for reflective or decorative purposes, as a vapor deposit. The metal is vaporized in a high vacuum chamber containing the parts to be coated. The metal vapor condenses on the exposed surfaces in a thin film that follows the surface pattern. The process is cheap for coating small parts, considering the time element only, but the cost of special equipment needed is relatively high.

Aluminum is the most used metal for deposit by this method and is used frequently for decorating or producing a mirror surface on plastics. The thin films usually require mechanical protection by covering with lacquer or some other coating material.

HOT DIP PLATING

Several metals, mainly zinc, tin, and lead, are applied to steel for corrosion protection by a hot dip process. Steel in sheet, rod, pipe, or fabricated form, properly cleansed and fluxed, is immersed in molten plating metal. As the work is withdrawn, the molten metal that adheres solidifies to form a protective coat. In some of the large mills, the application is made continuously to coil stock that is fed through the necessary baths and even finally inspected before being recoiled or cut into sheets.

Zinc Applied in Many Ways. Zinc is one of the most common materials applied to steel in this manner. In addition to protection by exclusion, electrochemical protection (the source of the term *galvanized iron*) occurs when exposed steel and adjacent zinc are connected by conducting moisture. Zinc is one of the most favored coatings for corrosion protection of steel because of its low cost and ease of application. In addition to hot dipping, zinc can also be applied by electroplating, spraying, and *sherodizing*. Sherodizing is a process by which steel, heated in the presence of zinc dust, becomes coated with zinc.

Tin plating and terne plating, the latter using a mixture of approximately four parts lead to one part tin, are also done by hot dipping.

ELECTROPLATING

Coatings of many metals can be deposited on other metals, and on nonmetals when suitably prepared, by electroplating. The objectives of plating are to provide protection against corrosion, to improve appearance, to establish wear- and abrasion-resistant surfaces, to add material for dimensional increase, and to serve as an intermediate step of multiple coating. Some of the most common metals deposited in this way are copper, nickel, chromium, cadmium, zinc, tin, silver, and gold. The majority are used to provide some kind of corrosion protection, but appearance also plays a strong part in their use.

Complex Electrical and Chemical System. Figure 16-5 is a schematic diagram of a simple plating setup. When direct-current power of high enough voltage is applied to two electrodes immersed in a water solution of metallic salt, current will flow through the circuit causing changes at the electrodes. At the negative electrode, or cathode (the work), excess electrons supplied from the power source neutralize positively charged metallic ions in the salt solution to cause dissolved metal to be deposited in the solid state. At the positive electrode, or anode (plating metal), metal goes into solution to replace that removed at the other electrode. The rate of deposition and the properties of the plated material are dependent on the metals being worked with, the current density, the solution temperature, and other factors.

Coating Thickness Usually Low. Thickness of plating is usually low, in the range of 2.5 microns to 0.025 millimeter (0.0001 to 0.001 inch). Chromium applied for appearance only may be used in a thickness of only about one-tenth these amounts, but when used to provide wear resistance and to build up dimensions, as on gages, may be applied in thickness as much as 0.25 millimeter (0.010 inch).

When plating thickness is a critical consideration, measurement and control may be established with NDT. Both eddy current methods and radiation back scatter are useful.

Multiple Metals for Maximum Properties. Layers of different metals are sometimes plated for maximum properties. For example, an object such as a steel bumper for an automobile may first be copper plated to provide good adhesion and coverage of the steel and to facilitate buffing to a smooth surface necessary for high quality final finish. Nickel is then plated over the copper to serve as the principal corrosion protection. Finally, chromium is plated over the nickel to serve as a hard, wear-resistant, bright,

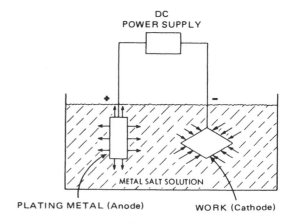

Figure 16-5
Electroplating

blue-white color coating over the softer, tarnishable nickel.

Many Problems Even Though a Common Process. Some problems exist with electroplating. Deposit on irregular shapes may vary widely in thickness. Projections and exposed surfaces may plate readily, but recesses, corners, and holes can sometimes be coated only by using specially located electrodes or electrodes shaped to conform to the workpiece shape. Electroplating can be costly because it involves payment for considerable electric power and the metal plated and lost. Because plating thicknesses are usually very small, the coating has little hiding power.

CHEMICAL CONVERSIONS

A relatively simple and often fully satisfactory method for protection from corrosion is by conversion of some of the surface material to a chemical composition that resists attack from the environment. These converted metal surfaces consist of relatively thin (seldom more than 0.025 millimeter, or 0.001 inch thick) inorganic films that are formed by chemical reaction with the base material. One important feature of the conversion process is that the coatings have little effect on the product dimensions. However, when severe conditions are to be encountered, the converted surface may be only partial protection, and coatings of entirely different types may be applied over them.

ANODIZING

Aluminum, magnesium, and zinc can be treated electrically in a suitable electrolyte to produce a corrosion-resistant oxide coating. The metal being treated is connected to the anode in the circuit, which provides the name *anodizing* for the process. Aluminum is commonly treated by anodizing that produces an oxide film thicker than, but similar to, that formed naturally with exposure to air. Anodizing of zinc has very limited use. The coating produced on magnesium is not as protective as that formed on aluminum but does provide some protective value and substantially increases protection when used in combination with paint coatings.

Purposely Created Oxide Better Than Naturally Formed Oxide. Because of their greater thickness and abrasion resistance, anodic films offer much better protection against corrosion and mechanical injury than do the thin natural films. Aluminum is usually treated in a sulfuric acid electrolyte that slowly dissolves the outside at the same time it is converting the base metal to produce a porous coating. The coating can be impregnated with various materials to improve corrosion resistance. It also serves as a good paint base and can be colored in itself by use of dyes.

The usual commercial anodizing methods used on aluminum cause formation of billions per square inch of aluminum oxide cells which grow above the original metal surface and at the same time extend below that original surface. Each of those cells has a pore in its center that extends to a solid barrier layer near the bottom of the cell as pictured in Figure 16-6. These numerous pores permit impregnation of the surface with various desirable materials but they are also a source of problems for penetrant testing of anodized aluminum surfaces. The penetrant can enter the pores to such an extent that an extremely high background is produced. Special care to interpretation of results may be necessary.

Checking for cracks is often called for because aluminum oxide is brittle and subject to cracking particularly if deformation of the material occurs after anodizing.

CHROMATE COATINGS

Zinc Dimensions Increase with Corrosion. Zinc is usually considered to have relatively good corrosion resistance. This is true when the exposure is to normal outdoor atmosphere where a relatively thin corrosion film forms. Contact with either highly aerated water films or immersion in stagnant water containing little oxygen causes even corrosion and pitting. The corrosion products of zinc are less dense than the base material so that heavy corrosion not only destroys the product appearance but also may cause malfunctions by binding moving parts.

Chromium Salts Improve Corrosion Resistance and Paintability. Corrosion of zinc can be substantially slowed by the production of chromium salts on its surface. The corrosion resistance of magnesium alloys can be increased by immersion or anodic treatment in acid baths containing dichromates. Chromate treatment of both zinc and magnesium improves corrosion resistance but is used also to improve adhesion of paint.

PHOSPHATE COATINGS

Used Mainly as a Paint Base. Phosphate coatings, used mostly on steel, result from a chemical reaction of phosphoric acid with the metal to form a nonmetallic coating that is essentially phosphate salts. The coating is produced by immersing small items or spraying large items with the phosphating solution. Phosphate surfaces may be used alone for corrosion resistance, but their most common application is as a base for paint coatings. Two of the most common application methods are called *parkerizing* and *bonderizing*.

CHEMICAL OXIDE COATINGS

A number of proprietary blacking processes, used mainly on steel, produce attractive black oxide coatings. Most of the processes involve the immersing of steel in a caustic soda solution, heated to about 150°C (300°F) and made strongly oxidizing by the addition of nitrites or nitrates. Corrosion resistance is rather poor unless improved by application of oil, lacquer, or wax. As in the case of most of the other chemical conversion procedures, this procedure also finds use as a base for paint finishes.

Inspection 17

Inspection is an essential procedure carried on in connection with all manufacturing processes by which usable goods are produced. Inspection work, however, differs from that of all the processes discussed to this point. Unlike them, it does not change the individual product, but instead, by elimination of bad parts improves the average quality of those that remain for distribution and use.

In general terms, inspection can be defined as an examination to determine the conformance of parts or assemblies to their specifications. The information gathered from such an examination may be used for several purposes. Because it is frequently impossible to manufacture articles within close enough limits that all can be used interchangeably, the inspection information is frequently used to sort products into groups. The information gathered from inspection is also used as an indication of need for adjustment of equipment or processes. A third objective of inspection procedures is to provide data for control of quality.

Quality Control Uses Inspection Data for Process Improvement. Although the term *quality control* is occasionally used synonymously with inspection, its meaning is sometimes different. The association between quality control and inspection is close. Quality control is often a second step, making use of inspection data for analysis and decision making for achieving, maintaining, and improving quality of products. In some manufacturing plants, both inspection and quality control are performed by the same department and personnel. In others, they are completely separated and may even have separate data collecting facilities.

INSPECTION PROCEDURES

Because of their effects on the product function, the selection of dimensions, qualities, and appearance factors for any product is primarily a design problem. In many cases the choices are empirical in nature, being based on past experiences, and in some cases are even arbitrary because of the lack of real information on which to base the kind of choice. Most dimensions and qualities are subject to wide variability in the manufacturing process and, in some cases, may also be very difficult to measure.

The desired life expectancy for a product also will usually play an important part in the consideration given to dimensions and qualities needed for satisfactory manufacturing. Because of these factors and the close association between processing and quality control, both the manufacturing and inspection divisions of a manufacturing plant are often consulted before a final determination of quality tolerances. In addition, they are usually the principal decision makers for setting the inspection qualities, quantities, and standards.

Inspection Varies with Quality Desired. The difference in the amount and kind of inspection necessary for a machine tool as compared to a piece of farm equipment is considerable. In the first case, a machine tool would be expected to be rigid, to be free working with a minimum of friction loss, to have very accurate related surfaces for maintenance of accurate movement, to have long life, and, during that period, to be able to produce parts accurate within a few ten-thousandths of an inch of dimension. These requirements mean that most of the parts of which the machine is constructed must be held within extremely close accuracy limits, and large amounts of inspection are necessary.

In the case of the farm machinery, which may be no less important in its own area, the product must be strong, rugged, able to withstand exposure to the elements, and also to function over a long period of time, although the actual hours of use may be relatively few. The farm machinery, however, does not require the relationship accuracies that must exist in the machine tool, so that both the quality and quantity of inspection can be reduced. These differences naturally show up in the cost of the completed equipment.

Inspection Benefits Management and Customer. The meeting of specifications set by the designer is primarily a manufacturing problem. Whether or not the specifications are met is determined by inspection, which may be performed by either operating or specialized personnel. Regardless of his other duties, an inspector at the time he is performing this function may be considered to represent both management and the customer.

Processing Closely Related to Quality. Any product is always subject to quality variation, the degree of which will vary in wide ranges depending largely upon the relationship between the product design and the process chosen for its manufacture. The materials of the product, the equipment used, the personnel operating the equipment, and the planned steps by which the manufacturing is carried on are all influencing factors on the quality variation. Inspection is for the purpose of finding these variations and, in many cases, aiding in assigning the causes for their existence.

ORGANIZATION OF INSPECTION

Inspection Always Present. Although certain kinds of inspection are limited to certain phases during the manufacturing processes, inspection of some type, sometimes as simple as casual observation, is needed in every stage of manufacturing of every kind of product. It is, however, customary in many plants to label in general terms the inspection procedures according to the state of the product being examined, as receiving inspection, in-process inspection, and final inspection.

Receiving Inspection. The term *receiving inspection* denotes all the inspections, regardless of type, that are given to incoming material, including such things as raw materials, speciality items, and subassemblies manufactured under subcontract. To cut down transportation and handling, companies making use of large quantities of speciality items or subcontract work frequently perform this kind of inspection in the supplier's plant.

In-Process Inspection. Inspection that is conducted during the time raw material is being converted into a finished product is called *in-process inspection*. The place of inspection is dependent largely upon the degree of examination and the kind of equipment needed. When only a percentage of the parts produced are inspected, either periodically or in spot checks, the work is usually carried on at the machine. Particularly in small plants, this inspection may be performed by the machine operator himself. When large quantities of product are to be inspected, and when the inspection procedures require specialized equipment, the work is most often done in centralized areas.

First-Piece Inspection Part of In-Process Inspection. Regardless of the amount of other inspection that might be necessary, *first-piece* inspection is common practice. After any equipment setup, tool change, or any action that may influence the quality of the product, the first piece is examined to determine its conformance to specification. This is sometimes a very formal procedure, and in many cases, as in pressworking where the effect of wear and other factors is small, this may be the only inspection required.

Final Inspection. Inspection performed at *final* inspection may include a great variety of work. Visual inspection for appearance (paint, labels, cleanliness) and completeness (all parts, instruction books, parts list) is nearly always part of the job. Tests for function, which are sometimes necessary on mechanical goods, may involve elaborate testing procedures requiring much time and adding considerable cost to the overall manufacturing operation. Testing of most aircraft in the final stages would fall in this category.

When the amount of final inspection is large, reduced in-process inspection may be permitted, although this will depend on a number of factors, including the relation of inspection cost to processing cost and the cost of replacing bad parts in the final assembly.

Nondestructive Testing. The vast majority of inspection performed on manufactured goods is nondestructive in nature but most measurements of dimensions, geometry, appearance, completeness, and the like do not fit the usual concept of NDT. NDT usually involves indirect tests that are in some way related to qualities and characteristics that cannot be checked directly without destruction. This kind of testing not only fits into all of the described areas of inspection, but is essential if there is to be assurance of good quality product.

QUANTITY OF INSPECTION

The percentage of inspection at any phase of manufacturing will vary widely. When lowest inspection cost is the principal interest, the variation can be from 0% to 100%. When greatest reliability is of interest, 0% would be unlikely, but 100% may also be unlikely because 100% inspection does not always mean 100% reliability due to the effects of fatigue and monotony as well as the psychological and hypnotic effects of continuous detailed work.

Desired and Experienced Quality Determine Quantity Inspected. With a large portion of manufactured goods, the quantity to be inspected is determined by the use of various sampling plans. These may be used only in those cases where something less than 100% perfect quality will be accepted. In general, the lot size being inspected must be large because of the assumption that the inspected quality will vary according to known statistical laws. Mathematical methods are available for designing a number of sampling plans that take into account the product quality level and the willingness to accept a certain defective part. The necessary sample size is affected by these factors.

Randomness of Sample Important. For any sampling plan to be effective the sample inspected must be random and truly represent the overall quality of the lot. Before a complete sampling plan can be devised, a decision must be made as to the percentage of defective parts in a lot that would be willingly accepted. Ideally, a sampling plan would accept all good lots and reject all bad lots of parts.

Most Economical Sample Size a Compromise. The ideal, however, can be reached only when the sample size becomes 100% and is, in addition, performed without fault. As shown in Figure 17-1, ideal results are approached when the sample size is increased; consequently the best sample size is always a compromise based on the relative values of improved reliability versus greater inspection costs. Acceptance sampling plans are essential when inspection cost is high and the cost for replacing defectives is low, when the sampling plan is more efficient than 100% inspection, and in every case when the inspection procedure is destructive.

Always Some Risk of Nonrepresentative Sample. The operating characteristic curve shown in Figure 17-2 is a single sampling plan requiring an attribute (quality that is either wholly present or absent) of

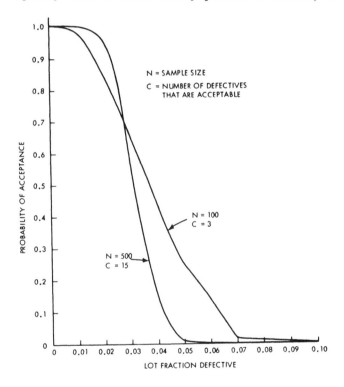

Figure 17-1
Operating characteristic curves for different sample sizes

200 randomly selected parts to be compared with its specification. If four or less defective parts are found in the sample, the entire lot from which it came will be accepted. If more than four defectives are found, the lot will be rejected and likely be sorted for removal of the defectives. In the plan shown, the dotted line marked P_1 indicates the so-called producers risk. If the lot being inspected had only 1% defectives, there would be a 6% chance that this plan would reject the material. The dotted line marked P_2 indicates the *consumer's risk*, which in this case is a 10%

chance that a lot with 4% defectives might be accepted. Sampling plans of this type therefore must be designed to be acceptable to both the producer and the consumer.

PROCESS CONTROL CHARTS

Need Variables Instead of Attributes. Another valuable use of statistical mathematics in inspection is for the construction of control charts with limit lines. Inspection values plotted on the chart will rarely fall outside these lines except when an assignable cause exists. In other words, the variation of points inside the control limits can be from chance causes alone. The data collected for construction of process control charts is in the form of variables rather than attributes. Data collection is therefore more costly, but in most cases considerably more information can be made available from analysis of the data.

Assumptions Do Not Destroy Value. In the making of control charts, some assumptions are

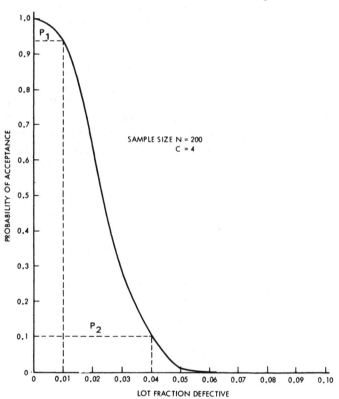

Figure 17-2
Operating characteristic curve

made, which, although they may not be entirely true, can usually be approximated closely enough that the system will work. One of the important assumptions is that variation of the quality being inspected will follow a known frequency distribution. Most often it is assumed that the frequency distribution follows a normal curve, as shown in Figure 17-3. In a normal distribution, 99.73% of the measured values from an entire population will probably fall within the limits of ±3σ from the arithmetical mean. (Sigma is the symbol for *standard deviation*, which is a measure of the dispersion of the measured values.) Similarly, 95.46% of measured values would be expected to fall within ±2σ limits, and 68.26% within ±1σ.

Chart Constructed from Process History. The construction of a quality control chart usually follows the following kind of procedure. First, the process is examined to ascertain that it is normal and that all assignable causes have been eliminated so that its operation is stable within the limits of chance variation. Next, an historical record is made by plotting, the mean values of a number of samples, the size frequency, and selection of which have been carefully predetermined after consideration of the process characteristics. These values are placed on two charts, one for averages and one for ranges, and limits calculated for each (Figure 17-4). If the limits used are ±3σ, not more than 0.2% of any plotted points would be expected to fall outside these lines. Therefore,

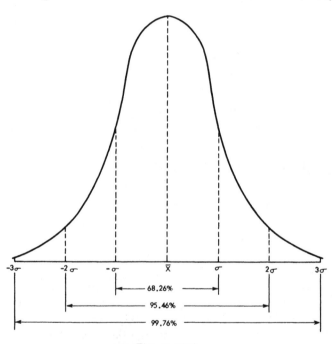

Figure 17-3
Distribution under a normal curve

whenever a point does fall outside, the process is critically examined for an assignable cause.

As the process continues, current samples are plotted and compared with past history to determine that the process remains in control. In most processes, the mean is controllable by adjustment of the process, but the range can be changed only by finding and eliminating assignable causes.

Charts Best for Long Runs. Although process control charts can be useful for short-run operations under some conditions, their greatest value is in continuing operations in which a minimum number of changes may contribute to variability. The information that can be gathered from control charts can be useful for several purposes. It may be used for

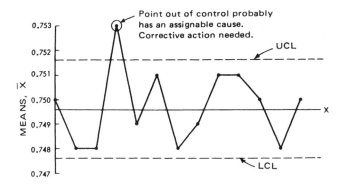
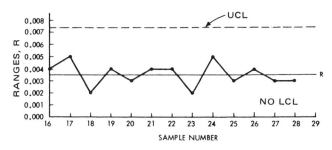

Figure 17-4
Quality control process chart

determining the overall quality of a product. The data can be useful for matching mating part dimensions with a minimum of waste. Understanding of the statistical variation in a product usually will permit wider tolerance use. Although all the points within the control limits on the mean chart could be in these positions by chance variation, a gradual shift toward one or the other limit can often be interpreted as a trend caused by an assignable reason. For example, gradual tool wear in a cutting operation would cause the average mean value to change gradually.

Process Improved by Identification of Causes. Frequently, the use of process control charts will cause improvement in the processes on which they are used by pointing out possibly correctable variation causes. Analysis of the process itself and correction of faults as they are found will produce gradual improvement in the process history and tend to tighten down on the control limits as they are recalculated. The presence of regularly kept charts in the process area tends to have a rather large psychological effect on the operators. Frequently they do a better job merely because the chart is before them. The data that are collected for construction of the control chart are, of course, useful also for inspection acceptance, and often provide more information than would be available from data collected for inspection alone.

PRINCIPLES OF MEASUREMENT

This section of inspection is concerned primarily with dimensions, shapes, finishes, dimensional tolerances, and the dimensional relationships, together with geometric relationships existing between surfaces. Any quality desired in a manufactured product may require inspection to assure its meeting specifications. In the manufacture of hard goods, the greatest amount of inspection time is spent checking those qualities mentioned in this paragraph. Some important properties such as hardness and strength, together with their testing procedures, have been discussed in earlier chapters.

DIMENSIONAL REFERENCES

Use of Common Reference Points Valuable. When dimensional measurements are being made, a reference point and a measured point always exist. In the case of single dimensions, it usually makes no difference which is which, except in those cases in which one surface is more rigid or more easily accessible and will serve better as a reference point. When a number of dimensions originate from the same point or can be measured from a common point, that point should be used as a reference point. All measurements should be made from it to reduce the possibilities of accumulation of error. When a series of dimensions are measured, each dependent upon the previous one, the total possible error is the accumulation of all the individual errors. But if, as shown in Figure 17-5, each measurement is made to a common reference point, the maximum total error can be only two individual errors for any of the dimensions measured.

In those cases where the only practical dimensioning method requires a sequential group of measurements, it is good practice to leave the least important dimension off the drawing and thereby eliminate the argument as to whether the overall dimension or

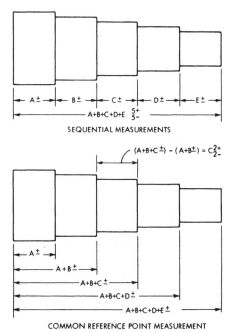

Figure 17-5
Accumulation of dimensional error

individual dimensions should receive first consideration regarding the holding of tolerances.

Drawings and Procedures Should Agree. Drawing dimensions should always agree as closely as possible with manufacturing and inspection procedures to minimize the need for calculations by machine operators and production personnel. When changes in a process cause changes in measurement procedures, action should be taken to correct the working drawings to fit the new methods.

TOLERANCES

Tolerances Should Fit Product and Process Need. Although it is possible by use of sufficient time and care to work as closely to a given dimension as is desired, it is impossible to manufacture to an exact size. Regardless of the accuracy displayed, it is always possible to choose a finer measuring method that can show discrepancies in the dimension. As working to higher accuracies costs more in money, time, and equipment, it is most economical and practical that dimensions should be permitted to vary within the widest limits for which they can still function properly. This variation is permitted by the use of *tolerances* added to dimensions in such a way that they indicate the permissible variation. Theoretically at least, the designer applies dimensional tolerances as wide as can be safely used. One of the inspector's jobs is to determine whether the product is made within these manufacturing limits.

Basic Dimensions Displayed as First Goal. Manufacturing tolerances may be shown in different ways, as indicated in Figure 17-6. If a dimension is approached in a definite direction by the manufacturing process used, and greater chance of error exists on one side of the basic dimension than on the other, unilateral tolerances are usually displayed, using the dimension that would be reached first as the basic dimension. When no reason exists for error on one side of the basic dimension more than on the other, bilateral tolerances permitting variation in both directions are used. The third method shows both limiting dimensions and thereby eliminates the need for calculation by production personnel. However, it tends to clutter up the drawings because of its sometimes greater space requirement and the increase of significant numbers.

Understood Tolerances — Local Agreements. The majority of dimensions on drawings are not critical and are usually shown without tolerances indicated. However, to prevent complete loss of control, these are usually treated to have *understood* tolerances that may vary in different plants but are usually in the range of ±0.010 to ±0.015 inch.

SOURCES OF MEASUREMENT VARIATION

Variation in dimensional measurement comes from a number of sources. Some are common enough that they should be given consideration in the majority of measuring and inspection procedures. Among these are parallax, temperature effects, pressure effects, and human error.

Parallax Is an Apparent Displacement. The illusion created by parallax is shown in Figure 17-7. If the hand swinging over the scale is viewed from Point A, directly in front, measurement 5 would be observed. If, however, the eye were moved to position B, the hand in the same position would indicate a reading of 6. This is the illusion that makes it difficult to read a clock correctly when viewing it from an angle.

Any measuring or indicating device that has a finite thickness between the indicating member and the reading scale or the work will display an error caused by parallax if used incorrectly. Many meters are constructed with mirrors underneath the indicating hand so that, to obtain a single view of the hand, the eye must be positioned in the only spot where a correct reading can be directly read. Many meters and instruments used for NDT are so equipped.

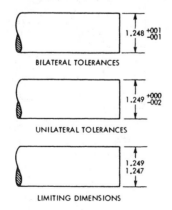

Figure 17-6
Methods of showing dimensional limits

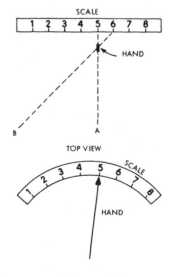

Figure 17-7
Parallax

Temperature Effects Often Present. It is well known that temperature variation causes changes of dimension in materials, causing them to grow larger with increased temperature and smaller with decreased temperature. Different materials are affected to different degrees by temperature changes or in other words, have different coefficients of thermal expansion. Many of the manufacturing processes cause temperature changes in the work and in the gaging and measuring equipment or are concerned with different materials such that measurement problems caused by temperature are significant.

The coefficient of thermal expansion for steel is approximately 0.0000117 unit per unit per °C (0.0000065 unit per unit per °F). It would not be unusual for a steel disc being machined to 150 millimeter (6-inch) diameter to have its temperature increased during the machining work to 120°C (200°F) above standard temperature of 20°C (68°F). If measured while still hot with a gage calibrated for use at standard temperature, an error of about 0.21 millimeter (0.008 inch) would be measurable on the disc when cooled to standard temperature.

$120 \times 1.17 \times 10^{-5} \times 150 = 0.21$ mm
$(200 \times 6.5 \times 10^{-6} \times 6 = 0.0078$ in.$)$

Aluminum, for which the coefficient of expansion is approximately 0.0000216 unit per unit per °C (0.000012 unit per unit per °F), would under the same conditions be expanded almost twice as much and upon cooling would show an error of more than 0.38 millimeter (0.014 inch).

$120 \times 21.6 \times 10^{-6} \times 150 = 0.389$ mm
$(200° \times 12 \times 10^{-6} \times 6 = 0.0144$ in.$)$

When using a steel measure or gage on a steel workpiece, little error would be caused if both were at the same temperature (dependent somewhat upon the gage design). However, in the case of the gage and the work being of different materials, such as a steel gage on an aluminum part, exact measurement can be made only when both are at standard temperature. For example, if the above aluminum disc and steel gage were both at only 20°C (36°F) above standard temperature, the error in measurement would be almost 0.03 millimeter or more than 0.001 inch.

$20 \times 21.6 \times 10^{-6} \times 150 = 0.0648$ mm
$20 \times 11.7 \times 10^{-6} \times 150 = 0.0351$ mm
$0.0648 - 0.0351 = 0.0297$ mm

$(36 \times 12 \times 10^{-6} \times 6 = 0.00259$ in.$)$
$(36 \times 6.5 \times 10^{-6} \times 6 = 0.00140$ in.$)$
$(0.00259 - 0.00140 = 0.00119$ in.$)$

Temperature also affects resistivity of material and changes flow of electric current. Therefore, eddy current results may be affected to the point that temperature readings should also be recorded when temperatures are different from normal. Non-uniform temperatures over a part being checked may also cause readings to vary in different locations when no difference in the tested attribute really exists.

For critical dimensions, particularly those of small size when the percent error will be large, care should be taken to see that the product being tested and any comparison standards are at the same temperature level.

Pressure Springs or Deforms Work and Equipment. For most dimensional measurement, some element of the measuring device must make contact with the work surfaces. The effect of the contact pressure depends on the strength and rigidity of both the work and the measuring tool and on the loads applied. Most measuring devices are constructed to use light pressures that only break through oil and dirt films on the surfaces, as contact is often only at a point or along a line until deformation causes sufficient bearing area to carry the applied load. It must be remembered that load can be carried only by a reaction of bending or deformation; consequently, light and repeatable contact pressures are a necessity to accurate dimensional measurement.

Human Element a Large Variable. One of the most difficult problems to deal with in inspection, as well as in all the other phases of manufacturing, is error caused by the human element. Inspection procedures making use of any of the human senses (sight, hearing, smell, taste, or touch) are subject to some variation with any individual and usually to large variation between individuals. Sight and touch in particular are frequently used as part of a measuring system. At any time great reliability is required, the procedure should be designed to minimize the effects of the human element.

BASIS FOR MEASUREMENT

Measurement of various attributes may be either comparative or absolute. In many cases knowledge of the value of a dimension or other quality is unimportant, and interest is focused on measurement of the difference from some standard.

Many kinds of gaging apparatus are designed to show only the nearness or farness of a measurement from a predetermined standard.

Other gaging equipment sets the limits within which a dimension must fall to be acceptable and also does not assign any real value to the measurement.

A third type of measurement provides knowledge regarding the real or absolute value of a measurement by comparing the measurement with a known standard.

Comparison with Standards May Be Converted to Absolute. The differential measurements described in the preceding paragraphs can be converted to absolute values by addition or subtraction of the reading with the standard if its absolute value is known. All absolute measurements use zero as a reference point.

Metric and English Measuring Systems. Two measuring systems are commonly used throughout the world. These are the metric and the English systems, with the metric being more widespread but the English being more important to manufacturing in the United States until the current time. The metric system is universally used in most scientific applications but, for manufacturing in the United States, has been limited to a few specialities, mostly items that are related in some way to products manufactured abroad.

The Metric System Soon to Be Worldwide. England is currently in the middle of an official change from the old system to a metric system similar to that used in most of the world. The United States is not as far along in a similar change to the international system of units, which is a simplified form of the metric system, but there is little doubt the change will continue and accelerate.

United States Changeover Beginning. Some primary schools in the United States are introducing the new system to students. A few factories have already changed to metric units, and others are studying the problems, both functional and economic, connected with the change. There are some incompatibilities to be ironed out, and there are bound to be difficulties for those familiar with the English system becoming comfortable with a replacement.

New System to Be Simpler to Use. The international system of units (SI) simplifies calculations because of the multiple of ten relationship. Although some measurements will eventually be performed completely with the new units, some will require a long period for the change, and all during the transition will require conversion at times. This text has been written with dual units to help with familiarization of the relationship, but an attempt has been made to emphasize the new system to encourage its use. As an aid to conversion, some tables showing the relationships between the two systems are available in the appendix.

Length Standard Definitions. Length measurement standards are essential in order that units of measure have any meaning. All length measurements are related to the standard meter, which at one time was the distance between two marks on gold buttons placed on a platinum-irridium bar stored in Paris, France. Since the year 1960, a standard meter has been defined as being 1,650,763.73 wavelengths of light emitted from krypton-86. In 1866, the Congress of the United States defined a legal yard as being 3600/3937 of the length of a meter. From this definition, 1 inch turns out to be slightly more than 25.4 millimeters. More recently, the inch has been defined as exactly 25.4 millimeters. The meter and the inch are therefore primary measurement standards to which all length measurements are related.

Length Measurement Standardized by Gage Blocks. The use of uniform length measurement throughout the country is made possible by the use of secondary standards in the form of gage blocks that are used in three ways. Master gage blocks, the most accurate obtainable (guaranteed to be accurate within ±0.000002 inch per inch of length), are used only for checking other gage block sets so that their accuracy may be retained. Other sets of gage blocks, which may be of less original accuracy, are used as references and inspection blocks for the manufacture, calibration, and setting of various measuring devices. A third use applies blocks directly to precision measuring work in shop operations. The more gage blocks are used, the more important it becomes that they be frequently checked against other blocks to detect inaccuracies from wear and abuse.

Various Size and Quality Sets. Gage blocks may be obtained in sets containing as few as five to more than one hundred individual blocks. They are used by selecting blocks of such size as needed and wrung together to make up a desired dimension. Wringing in this case implies the use of a twisting sliding motion between the blocks that places their extremely flat and smooth faces so close together that they adhere to each other and can be built up to larger dimensions without inaccuracy caused by added space between the contacts.

Special Gages and Masters for Production Control. A tertiary measuring standard is used in manufacturing in the form of gages and measuring devices designed for specific purposes, and in the form of master work parts that can be used for comparative measurements.

INSPECTION EQUIPMENT

The equipment to be described in this section is primarily for dimensional measurement. It employs some type of comparison, with the principal difference being in the degree of reference to an absolute standard. The steel rule, for example, has a built-in reference to zero. A dial indicator has no built-in reference and is used mostly for differential measurements, but it can be used for absolute measurement by establishing proper reference. The spring caliper may be used as a gage to establish a dimensional limit, or it can be used to transfer a dimension from a work surface to some measuring device. Measuring tools may be classified as direct-reading devices, comparators, or limit gages.

Direct-reading devices provide the widest range of measurement of any of the measuring tools but are slower to use than the other types. In general, they require greater skill from the user and are therefore more subject to human error.

Steel Rules for Relatively Rough Measurement. Among the most common of the direct-reading inspection devices are steel rules and their variations. Steel rules are made in all sizes, from ones a fraction of an inch long that must be held in special holders, up to those several feet in length. They may be calibrated in different ways, depending on the use for which they were intended, and sometimes are calibrated with four different scales on the same rule. Most common for use in the United States are calibrations showing 1/64, 1/32, 1/16, and 1/8 inch, although in some applications, divisions in hundredths are of value. Steel rules showing combinations of English and metric units or all metric units are also available.

Good quality steel rules are machine divided with the calibration marks accurately placed, but ordinarily cannot be expected to be used with accuracies closer than about ±0.5 millimeter or ±1/64 inch.

Variations of the Steel Rule for Improved Accuracy. The steel rule has a number of variations, including the hooked rule that can be held over a corner, caliper rules that have a fixed and a sliding jaw to permit setting and easier reading, and depth rules that can reach into recesses. Some of these rules are shown in Figure 17-8.

Vernier Caliper and Height Gage Similar. Vernier calipers are variations of the steel rule that can be

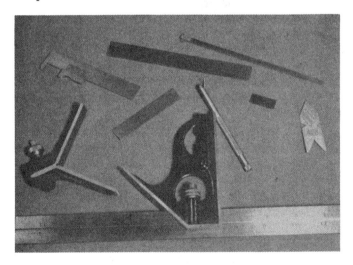

Figure 17-8
Steel rules

read to thousandths of an inch by use of a vernier scale built as part of the instrument. The height gage is similar to the vernier caliper with the exception that it is mounted on a base to hold it in a position suitable for vertical measurement.

Vernier Scales Are All Similar in Principle. Both instruments are calibrated as shown in the insert of Figure 17-9, with the main scale divided into inches and subdivided into 1/10 and 1/40 (0.025) inch. The vernier scale, which slides along adjacent to the main scale, has twenty-five divisions in the length equal to twenty-four divisions of the main scale and furnishes the witness line for reading a measurement. Each division on the vernier scale is 0.001 inch shorter than the similar divisions on the main scale, so that for each 0.001 inch of movement between the two, a different line on the vernier scale will line up with one of the marks on the main scale. A measurement reading is accomplished by first reading the full inches, adding tenths of an inch exposed before the zero of the vernier scale, adding 0.025 inch for each exposed subdivision, and finally adding the number indicated by the mark on the vernier that is in closest alignment with one of the marks on the main scale.

MICROMETER CALIPER

Micrometer Nomenclature. The micrometer caliper, or "mike," shown in Figure 17-10 is one of the most common measuring instruments used in the manufacturing field. For a precision tool, its construction is relatively simple. A U-shaped frame supports a hardened steel button called an *anvil* on the inside of one end and a *sleeve*, *barrel* or *hub* containing a threaded nut on the opposite end. The threaded nut supports threads on a spindle that extends through the sleeve and frame so that its flat end can be paired with the anvil to serve as the measuring element. The opposite end of the spindle is

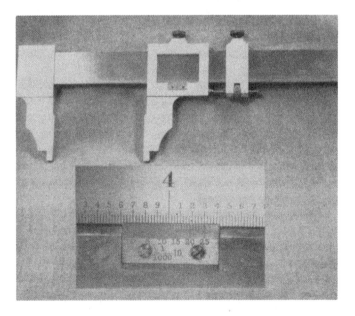

Figure 17-9
Vernier caliper

attached to a tubular *thimble* that rides over the outside of the sleeve so that when the thimble is turned, the spindle thread rotates in the fixed nut and causes the distance between the spindle and the anvil to decrease or increase.

Reading Is a Systematic Procedure. The threads of micrometers are the real measuring elements and are precision made, usually being ground in hardened materials. Forty threads per inch cause the thread lead to be 1/40, or 0.025 inch. A witness line along the side of the micrometer sleeve is divided into ten numbered divisions, each representing four full turns of the micrometer thread, a distance of 0.100 inch. Each 1/10-inch division is subdivided into four smaller divisions, each representing one full turn, or 0.025 inch. The bevel of the micrometer thimble is divided into twenty-five equal spaces to enable the user to read fractional turns with the accuracy permitted by 0.001-inch calibration.

Vernier Use Requires Careful Setting. Some micrometers also carry a vernier calibration consisting of ten marked spaces on the sleeve of the micrometer in a space equal to nine 0.001-inch divisions on the thimble. The principle of the vernier is the same as that on the vernier caliper and, with proper use, allows the micrometer to be read accurately to the nearest 0.0001 inch. Vernier micrometers calibrated to this accuracy are not too commonly used, however, because variations in temperature, pressure, and the human element frequently cause errors large enough to make this kind of accuracy impractical.

Frame Sizes Varied to Cover Large Range. Most micrometer heads are substantially the same in design and cover a 1-inch range. To permit wide range measurement, the heads are fitted to frames different in size by 1-inch increments. The tool is in common enough use in small sizes that the 1-, 2- and 3-inch micrometers (maximum limits) are usually personal tools of machine operators and mechanics.

Large Mikes Difficult to Use. Larger sizes, usually up to 24 inches, although larger than this have been built, are normally supplied from a company tool crib when their use is required. The larger sizes are naturally more difficult to position on work and to adjust with the correct "feel." Thus, frequently, some other device will be used when long dimensions must be accurately measured.

Mikes are rugged tools and can stand some abuse but should be accorded the careful treatment due a precision instrument. With relative ease, they can be used for measuring to accuracies of 0.001 inch; in the case of vernier mikes, they approach 0.0001 inch if proper consideration is given to temperature and pressure effects.

Other Applications. In addition to the outside micrometer described, the same principles are applied in the making of inside micrometers and depth micrometers for measurements and of various types of positioning screws for accurate locating-type applications. A bench-type *supermicrometer* is sometimes

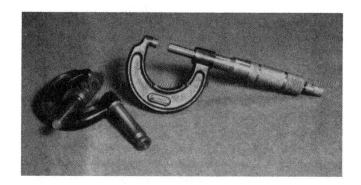

Figure 17-10
Micrometer

used in laboratories and tool rooms for accurate length measurements. This instrument also uses a screw thread for measurement but is constructed with a heavy frame consisting of a steel bar more than 3.5 inches in diameter, and incorporating spring loading on the workpiece so that very accurate measuring or contact pressure can be duplicated. The design eliminates practically all effects of the human element.

OTHER ADJUSTABLE TOOLS

Some commonly used adjustable inspection tools can be set to be used as limit gages but are more commonly used as dimension transfer devices. Inside calipers have turned-out legs to make contact with inside shoulders and holes. Outside calipers have turned-in legs for checking across the outside of shoulders or diameters of bar material. Hermaphrodite calipers, consisting of an inside caliper leg combined with a pointed divider leg, are primarily layout tools rather than measuring devices. Telescoping gages are made up of sleeves that can be locked in position to carry an inside dimension such as a hole diameter to a measuring device such as a micrometer.

Sine Bars or Tables for Accurate Angle Measurement. Angles may be measured in a number of ways, but one of the more precise methods used primarily in the laboratory and tool room is by use of a *sine bar*, illustrated in Figure 17-11. Sine bars are con-

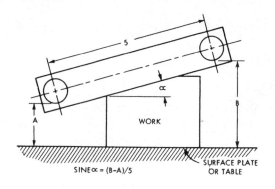

Figure 17-11
Sine bar

structed with accurately ground round buttons either 5 or 10 inches apart. The bar can be positioned with an angular position to match a workpiece. The difference in button heights from the base plane, divided by five for the 5-inch bar and by ten for the 10-inch bar, provides a number that is the sine of the angle of the bar's position in relation to the base. Accurate measurement of the button height is frequently performed by use of gage blocks.

INDICATING GAGES AND COMPARATORS

A second type of inspection device is the indicating gage or comparator, which is used for showing deviation from a dimension. By relating the reading to a suitable reference, these gages can provide absolute measure values. These devices require more skill than the direct-reading instruments for setup. Once set up, they may be used faster, easier, and frequently with greater accuracy than the direct type. Many also have the advantage of being combinable for multiple measurements and thus provide even greater time-savings. Most do, however, have a narrow measuring range for any single setup.

Most indicating gages and comparators are quite sensitive, with high amplification characteristics that may be provided by mechanical, electrical, pneumatic, or optical systems. They are used for comparing with known dimensions and with master workpieces and for checking parallelism, concentricity, and general conformance to shape.

Dial Indicators Have Many Applications. The majority of mechanical-type comparators are of the dial indicator style shown in Figure 17-12. These are constructed with a spindle that operates a rack gear in contact with a system of gears which turn the indicating hand over a calibrated dial. The use of light return springs to keep backlash from contributing error and of high quality bearing supports provide sensitivity permitting the indicator to be read accurately to within 1/10,000 inch. The majority of dial indicators are calibrated in thousandths of inches, but many, particularly in the larger diameters where the calibration marks can be better separated, are calibrated in 1/10,000 inch. The majority of dial indicators operate over ranges from about 1/16 to 1/8 inch, but some long range types have been designed to cover as much as 1 inch. These are constructed with an additional hand to count the multiple revolutions of the main indicating hand.

The majority of dial indicators are used for measuring dimension differences without regard to absolute values. Many special-purpose structures have been designed for supporting dial indicators for different kinds of uses. Some are special attachments designed to permit contact to be made with a surface difficult to reach. Some support a dial indicator in such a way that it may be used for work that would ordinarily be done with a fixed gage. Others hold the indicator so that it can be adjusted over a table where it can be used for making accurate comparison measurement.

No Joint Losses in Reed Mechanisms. The *reed* mechanism shown in Figure 17-13 is another method for amplifying small motions. One make of comparator gage uses this type of mechanism to move a small mirror. A light beam reflected by this mirror to a calibrated scale is in effect a weightless lever that increases the amplification of motion and provides extremely high sensitivity and response, permitting accurate readings in the range of 0.25 micron (1/100,000 inch).

Electrical Gages Permit Close Measurement. Electrical power is used for operation of both compara-

Figure 17-12
Dial indicator type of snap gage

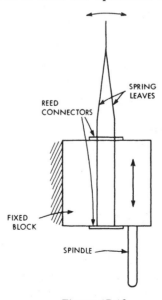

Figure 17-13
Reed mechanism for movement application

tor-type gages and limit gages. In the comparator type, movement of the work contact point of the gage from its zero or set position produces unbalance in the electrical system that causes current flow which can be read on a meter calibrated as finely as 0.025 micron (1/1,000,000 inch).

The electrical limit type of gage operates by the action of extremely sensitive switches that may be preset to definite dimensions. The switches may then be connected to operate signal lights, buzzers, or controls of gates in high speed sorting operations.

Pneumatic Gages Allow Noncontact Measurement. Air gages for making comparative dimensional measurements are of two types. In the pressure type, a pressure sensing element indicates a dimensional value on a calibrated scale as a result of back pressure built up from restriction of air flow through the gaging head. In the flow type, an indicator button floats on a column of air in a tapered glass tube, as air at constant pressure flows through a flexible tube and out orifices in the gage head. The gages are usually set with master workpieces or with limit gages that determine the limiting acceptance points.

Air gages are made with different degrees of amplification and sensitivity. Although they are used primarily as limit gages, a strong indication of absolute value is provided by the position of the indicator. Because air gage heads have some clearance with the surfaces they are designed to measure, their life is quite long. They are especially satisfactory for measuring materials that have abrasive characteristics or for use around abrasive processes such as grinding, honing, and lapping.

Optical Comparators Provide Enlarged View of Work. Optical comparators are designed to show a reflected surface picture, or a profile image, of a workpiece on a frosted glass screen. This is accomplished by casting light against the surface of the specimen and projecting its reflection through a magnifying lens system onto a mirror, which in turn reflects the image to the glass screen, or by passing light past the edge of the work to show its silhouette or contour (Figure 17-14). Most comparators permit lens changing to vary the magnification from 10 power to 100 power.

The enlarged image on the screen can be measured, observed visually for defects, or compared with enlarged drawings, frequently complete with limiting outlines, for inspection purposes. The equipment is especially useful for inspection of small, complicated shapes that would be difficult to examine carefully or measure by other means. Multiple dimensions and complex shapes can be quickly checked with this device.

Optical Flats Used for Flatness or Length Measurement. Another method of optical inspection involves the use of optical flats. These are flat, clear

Figure 17-14
Optical comparator projecting a magnified view of the work silhouette on a ground glass screen. Suitable for dimensional measurement as well as for checking shapes and relationships

discs, usually made of fused quartz, constructed with the two sides as parallel as possible. The principle upon which use of the optical flat is based is interferometry, a word used to indicate light wave interference to produce identifiable light and dark bands, as illustrated in Figure 17-15. Light waves from a monochromatic (single wavelength) light source are transmitted through the optical flat, which is set at a slight angle on the work surface. Part of the light will be reflected from the lower surface of the optical flat. Another part will pass this surface and continue on to be reflected from the work surface to rejoin the first part as both are reflected toward the observer's eye.

Depending on the distances each set of waves travel, some will be in phase and reinforce each other to form bright lines, while others will be out of phase, will interfere, and will cancel each other to produce

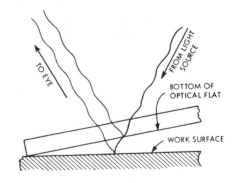

Figure 17-15
Light wave interference

dark bands or "fringes." Interference to form dark fringes will occur as the thickness of the air wedge between the optical flat and the work surface varies by one-half wavelengths. The frequency of bands will therefore depend upon the angle of the flat and the wavelengths of the light being used.

Optical flats may be used for checking flatness of surfaces because any deviation of the work surface being checked from the lower surface of the optical flat results in a pattern of fringe bands. The shape and spacing of the bands can be used to calculate accurately the degree of difference between the surfaces. Optical flats can also be used for making measurements as illustrated in Figure 17-16. In this case, 1-inch working gage block A is being compared with 1-inch master block B. Observation of the fringe bands of block A in the top view shows three complete bands indicating that if a monochromatic light source with a one-half wavelength of 11.6 microinches (0.0000116 inch) is being used, the optical flat is 3 x 11.6 or 34.8 microinches higher on one edge than on the other. By simple proportions, it can then be calculated that block A is shorter than block B by 3 x 34.8 or 104.4 microinches, and the height of block A is 1.000000 − 0.000104 or 0.999896 inch.

FIXED GAGES

A third type of inspection tool is the fixed gage,

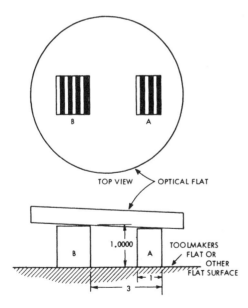

Figure 17-16
Work measurement with an optical flat

which is set to a limit of a dimension to establish a maximum or minimum value or to both limits to enclose the tolerance range. This type of gage measures attributes only and provides very little information regarding absolute measurement. Fixed gages are fast to use and require little skill to produce satisfactory results; consequently they are frequently used as production gages.

Most Are Special Production Gages. Most fixed gages are single-purpose tools, useful only for the dimension for which they have been set, although some of the standard types are adjustable and can be changed for other dimensions within a limited range. Fixed gages may be designed to check dimensions, shapes, relationships, or, in some cases, combinations of these qualities.

Pictured in Figure 17-17 are some typical fixed gages. Some, such as the plug gages and ring gages, are *go-not go* gages that are made with the two tolerance limits. Others, such as profile gages, are a negative shape of the part to be checked and may or may not be made to both tolerance limits, depending mainly on the importance of the shape and size. Progressive gages, such as a sequential series of increasing diameter plugs, come close to providing an absolute measure by dividing the overall tolerance into a number of smaller increments, thus tying the dimension down to a small range.

Figure 17-17
Fixed gages

Fixed Gage Tolerances Reduce Working Range. Gages, like any other manufactured articles, must be made to tolerances permitting some dimensional variation. These tolerances, naturally, must be smaller than the tolerances for the manufactured part on which the gage is to be used, and are usually held to between 10% and 20% of the part tolerance.

Gages must also be designed with some wear allowance so that they will not accept bad parts after a short period of use. The wear allowance used is variable, depending on the conditions of gage use, the precision of the product being inspected, and the gage life desired. In large operations, it is common for two sets of gages to be used. One set, called *working gages*, is made to the above tolerances and is used by the machine operators to check the product as it is being manufactured. The other set, *inspection gages*, are made to approximately one-half this tolerance, to reduce the chance of rejecting good parts.

Where many gages of the same type are used, master gages are sometimes constructed with tolerances 10% of the working gage tolerance for checking the gages themselves.

SURFACE FINISH

In addition to conformance to a general geometric pattern, many applications require that a surface have high quality finish.

Surface Variations of Different Frequency and Type. Three kinds of irregularities may occur on a surface. The one most evident is *roughness*, a term used to describe surface irregularities that are relatively close together. Surface roughness is usually a result of machining or other processing procedure that produces finely spaced irregularities.

A second surface fault is *waviness*, which refers to irregularities of wider spacing than those termed *roughness*. Waviness may be the result of warping, deflection, or springing while the workpiece is being worked upon, or the result of a tool movement pattern while the workpiece is being cut.

The third fault is an irregularlity called a flaw or imperfection, which is relatively infrequent and usually randomly located. Flaws consist of such things as scratches, holes, ridges, and cracks.

Surface Quality May Affect Function. The quality of some surfaces can play an important part in their function. Both flat and rotating bearing surfaces must usually be relatively smooth to function properly and often have their maximum roughness quality specified on their drawings.

Surface Marks Affect Fatigue Strength. Materials that are likely to be highly stressed in service, particularly by repeated or reversed load applications, may need good quality surface finish to reduce chances for fatigue failure. Any surface irregularity or discontinuity may be a point of stress concentration that can serve as a source of fatigue failure. As a precaution, the highly polished wing surfaces for high performance aircraft are frequently covered with a plastic coating for protection against nicks and scratches during manufacture because any marks on the surface might be a source of wing failure during flight.

Appearance Important to Saleability. The effect of surface finish on appearance alone should not be discounted. It is often the case that appearance is the only factor available for making a decision as to whether or not to purchase a product. It should be noted, however, that finish quality and light reflective ability are not necessarily synonymous. A newly finished clean surface with small, regularly spaced tool marks, particularly those made in a grinding process, will reflect light to produce a polished appearance. A random pattern of even smaller tool marks, such as might be made in a superfinishing operation, will not reflect light as well but will measure better, although appearing to be of lower quality finish.

Finish and Dimensions Closely Related. A close relationship exists between surface finish and linear measurement. Most measuring procedures involve the use of tools or instruments that physically contact the work surface and touch only on the high spots or peaks. A bearing surface might lose these peaks very quickly in use, and the large change of dimension that would occur with a rough surface would cause the original measurement to be meaningless. Good surface finish is certainly called for whenever close tolerances are required.

SURFACE FINISH MEASUREMENT

The roughness of a surface is made up of two qualities — the height and depth of irregularities, and the spacing between these. Most measurement methods take both into consideration to some degree without actually defining their relationship.

Lay — the Direction of the Principal Marks or Scratches. Most surfaces also will show different roughness measurements and characteristics in different directions. Measurements across the *lay* will in general be much higher than those with the lay. Lay is the direction of the predominant surface pattern. For example, a measurement across the lay on a piece turned in a lathe would be taken parallel to the workpiece axis.

Surface Comparison by a Variety of Methods. Some surface quality measurements depend upon comparison with standard samples displaying measured and known roughness. Visual comparison is sometimes satisfactory but often may not be too accurate because of the effect of dirt, corrosion, and irregularity of pattern on appearance. Accuracy of the comparison can be considerably improved by scraping a fingernail across the surfaces, adding a sense of feel. A visual method of comparing optical projection through a plastic film that has been pressed against the surfaces is also available. A film softened by solvent takes on the surface irregularities and by its refraction effect on the projected light rays causes a third-dimension effect on the screen, making accurate comparisons possible.

Electrical Instruments Most Common. The majority of accurate surface quality measurements are made with instruments that trace the work surface with a stylus, which in traveling over the hills and valleys disturbs an electrical circuit to make a reading possible. With some instruments, a pen is actuated to draw a magnified profile of the surface on a moving tape, in addition to a meter reading showing the average value of the surface traced. Other instruments show only the meter reading.

SURFACE SPECIFICATION

The specification of surface quality is indicated on the drawing, as shown in Figure 17-18. A 60° check mark is usually placed on the surface to which it refers, although in some cases, it may be located on a witness line, or an arrow may be used to indicate the

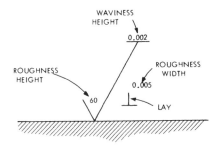

Figure 17-18
Drafting symbol for surface quality

surface. A number representing the maximum permissible roughness is located inside the V of the check mark. On the right side of the check mark is the lay symbol indicating the direction in which measurement should be made and the width of the maximum permissible roughness. Waviness is shown above a horizontal crossbar on the check mark.

On the drawing should be a note indicating whether the roughness value is total height, average height, or average deviation from the mean, either arithmetical or root mean square (RMS). It is most common to show only maximum value figures, although lower limits also may be indicated whenever they are of value.

Following are the lay symbols used to indicate a direction of measurement for which the figures in the specification apply:

= —Parallel to the boundary line of the nominal surface
⊥ —Perpendicular to the boundary line of the nominal surface
X —Angular in both directions to the boundary line of the nominal surface
M —Multidirectional
C —Approximately circular relative to the center
R —Approximately radial relative to the center of the nominal surface

Index

Abrasives, 150-51, 175, 176
  aluminum oxide, 151
  boron cubic nitride, 151
  diamond, 151
  silicon carbide, 151
Adhesive bonding, 157
  adherence, 161
  adhesives, 161
Adhesive joining, 95
Age hardening, 37
AISI numbers, 54-55
Allotropic charges, 37-38
Alloy steels, 51
  low alloy AISI, 53
  low alloy structured, 53
Alloys, 45
  aluminum, 58-60
  cobalt, 65
  copper, 60-62
  eutectic, 82
  magnesium, 64
  nickel, 62
  noneutectic, 82
  zinc, 64
Alumina, 163
Aluminum, 37, 58-59
  castings, 60
  heat treatment, 58
  pure, 58
  temper and heat-treat symbols, 60
  wrought, 58
Aluminum alloys, 58-59
  composition, 61
  electrical conductivity, 61
  general properties, 58-59
  properties, 61
  uses, 61
Aluminum-boron composites, 162
American Society for Metals *Metals Handbook*, Vol. II, 3
American Society of Nondestructive Testing, 3, 120
  Nondestructive Testing Handbook, 3
  Recommended Practice No. SNT-TC-1A, 2
American Society for Testing and Materials, 3
American Welding Society, 3, 117
Amorphous, 16
Annealing, 38, 39, 100, 138
Anode, 44
Arc cutting, 169
Arc welding, 108

Atomic structure, 15-16, 31
Austempering, 40
Austenite, 39
Austenitic stainless steel, 53
Austenitization, 38-39, 100
Autoclave, 163

Bainite, 40
Bar, 129, 130
Basic oxygen steel, 49
Bend testing, 25
  free bond test, 25
  guided bend test, 25
Bending, 18, 143-44
  forming, 144
Bessemer steel, 49
Billets, 128, 132
Blacksmithing, 134
Blast furnace, 46
Blasting, 175
  abrasive, 175
  liquid honing, 175
  sand, 175
  shot, 175
  shotpeening, 175
  steel grit, 175
Blooming mill, 129
Blooms, 128
Body-centered cubic lattice, 32, 33
Bonding, 96
  atomic, 96
  cold, 99
  flow, 97-99
  fusion, 96-97
  pressure, 97
Bonds
  atomic, 136
  mechanical, 136
Boring, 151
Brass, 62
  composition, 63
  definition, 62
  properties, 63
  uses, 63
Braze welding, 98
Breaking strength, 20
Brinell test, 26
Bronze, 62
  composition, 63
  definition, 62
  properties, 63
Buffing, 176

Carbon steel, 50
Carburizing, 138, 172-73
  carbon diffusion, 172
  grain-size control, 173
  casehardening, 172-73
  carburizing, 172
  flame hardening, 173
Cast aluminum, 60
Cast iron, 47, 48
  chilled, 48
  ductile, 48
  gray, 48
  malleable, 48
  white, 48
Cast steels, 54
Casting, 79-94, 148
  centrifugal, 93
  chaplets, 89
  chills, 86
  cold shots, 85
  cold shut, 85
  continuous, 93-94
  cores, 89
  crystal growth, 84
  design, 84-84
  die, 91
  flasks, 88
  foundries, 80
  gating, 85
  hot spots, 84
  investment, 92
  melting equipment, 94
  microshrinkage, 83
  mold, 80
  patterns, 80, 88
  permanent, 91
  plaster mold, 93
  plastics, 159
  porosity, 83
  pouring, 85
  process, 80
  risers, 86
  sand compaction, 88-89
  sand inclusions, 85
  sand molding, 86-90
  shrinkage, 82-84
Cemented carbides, 139
Cementite, 39
Centrifugal casting, 93
  centrifuge, 93
  semicentrifugal, 93
Ceramic coatings, 179
Ceramics, 163

Chaplets, 89
Charpy test, 24-25
Chemical milling, 166-67
Chemical oxide coatings, 181
Chills, 86
Chromate coatings, 181
Cleaning, 173
   alkalines, 174
   buffing, 176
   electropolishing, 176
   mechanical, 174-75
   pickling, 174
   polishing, 176
   solvents, 174
   vapor degreasing, 174
   water, 174
   wire brushing, 175-76
Coatings, 176-80
   anodizing, 179, 180-81
   brushing, 178
   ceramics, 179
   chemical conversions, 180
   chromate, 181
   corrosion protection, 176
   dipping, 178
   drying, 178
   effects on friction, 177
   electroplating, 180
   electrostatic spraying, 178
   enamel, 177
   galvanizing, 179
   lacquer, 178
   metal spraying, 179
   metallizing, 179
   oxide, 181
   paint, 177
   phosphate, 181
   plating, 177, 179
   porcelain, 178
   preparation for, 177
   spraying, 178
   vacuum metallizing, 179
   varnish, 177
   vitreous enamels, 179
Cobalt alloys, 65
Coining, 138
Cold finishing, 129-30
Cold pressing, 137
Cold shut, 85
Cold work, 35, 122, 125-26, 129
Collapsability, 89
Columnar grains, 82
Composite materials, 139, 162
   aluminum-boron, 162
   fiberglass, 162
   graphite, 162
   honeycomb, 162-63
   laminates, 162
Composites, 162
Compression testing, 21
Computers, 156
Concentration cells, 42
Concrete, 163
Continuous casting, 93-93, 129
Control charts, 186
Copper, 62
   corrosion resistance, 62
Copper alloys, 60-62
   brass, 62
   bronze, 62

   general properties, 61-62
Cores, 89
Corrosion, 4-5, 40-44
   atmospheric, 42
   definition, 40
   direct chemical action, 40-41
   electrochemical, 41-42
   fretting, 43
   general, 42
   intercrystalline, 43
   pitting, 43
   rate, 42
   season cracking, 43
   stress, 43
Corrosion protection, 43-44, 174
   chemical compounds, 44
   metal coatings, 44
   nonmetallic coatings, 44
Corrosion resistance, 116
Creep, 4, 24
Creep strength, 24
Creep testing, 24
Crucible, 94
Crucible steel, 48
Crystal growth, 33, 81
Crystal lattices, 32
   body-centered cubic, 32, 33
   face-centered cubic, 32, 33, 124
   hexagonal closed-packed, 32, 33
Cupola, 94
Cupping, 132
Cutting tools, 150-51
   carbon tool steel, 150
   cast nonferrous alloys, 150
   cemented carbides, 150
   ceramics or cermets, 150
   coated tool materials, 150
   diamonds, 150
   high speed steel, 150

Decarburization, 126
Defects, 5, 116-20, 129
   elongated, 123
   porosity, 137
Deformation, 122-25, 127, 128, 133, 136-37, 141
Dendrites, 81
Dendritic microporosity, 125
Die casting, 64, 91-92
   cold chamber, 91
   hot chamber, 91
Die set, 142
Dies, 126, 130, 132, 133, 134, 135, 136, 137, 142
   graphite, 137
Diffusion welding, 113
Direction effects, 122-23
Discontinuities, 5
Drawbench, 130
Drawing, 40, 135, 144
   shell, 144
Drilling, 151
Ductility, 21, 125, 126, 128, 141, 143
Dynamic loads, 4

Elastic deformation, 19
Elastic failure, 123
Elastic limit, 4, 34, 36, 122, 125, 142
Elastomers, 163

Electric furnace steel, 49
Electrical discharge machining, 164-66
   electrodes, 164, 165
Electrochemical machining, 166
   electrolytes, 166
Electroforming, 168
Electrolytic (electrochemical) reaction, 41-42
Electromagnetic forming, 146
Electron-beam welding, 111
Electroplating, 180
Electropolishing, 176
Electroslag welding, 112
Enamel, 177
Endurance limit, 23-24
Engineering materials, 9
English measurement system, 190
Equiaxed grains, 82
Eutections, 82
Explosion welding, 112-13
Explosive forming, 145
Extrusion, 132-33
   nonferrous materials, 133
   plastics, 159

Face-centered cubic lattice, 32-33
Facing, 151
Fatigue failure, 23, 125
Fatigue strength, 24, 196
Fatigue testing, 23-24
Ferrite, 39
Ferritic stainless steel, 53
Ferrous materials, 46-47
   ore reduction, 46-47
Fiberglass composites, 162
Filler, 97
Finish, 154
   imperfections, 154
   lay, 154
   roughness, 154
   waviness, 154
Finishing
   abrasive barrel, 175
   blasting, 175
   buffing, 176
   electropolishing, 176
   polishing, 176
Flame hardening, 173
Flasks, 88
   cope, 88
   drag, 88
Flow growth, 4, 6, 7
Flow rate, 122
Forging, 133-39
   blacksmith, 134
   closed die, 133, 134
   drop, 135
   hammer, 134
   hand, 134
   manual, 134
   open die, 133, 134
   press, 135
   roll, 135
Forming, 144
   electromagnetic, 146
   explosive, 145
   high energy rate, 145
   inductive-repulsive, 146
   roll, 144
   stretch, 144

Foundries, 80
Foundry mechanization, 94
Fracture control, 6-8
    nondestructive testing, 6-7
Fracture failure, 123
Fracture mechanics, 7
    critical plain strain stress intensity, 7
    fracture toughness, 7
    plane strain, 7
    stress intensity, 7
Fretting, 43
Friction, 177
Friction sawing, 169
Friction welding, 112

Gage blocks, 190
Galvanic cells, 42
Galvanic series, 40-41
Gas metal-arc welding, 109
Gas tungsten-arc welding, 108
Gas tungsten wire welding, 109
Grain boundaries, 33
Grain growth, 124
Grain size, 33, 124-25
Grain-size control, 36, 38
Graphite composites, 162
Green sand, 87
Grinding, 151

Hammer forging, 129
Hardening, 138
Hardness testing, 25-28
    Brinell, 26
    File test, 26
    Knoop, 28
    microhardness, 28
    Mohs test, 25-26
    Rockwell, 26-27
    superficial, 27
    Vickers, 27
Heat- and corrosion-resistant alloys, 64-65
Heat treatment, 138
    aluminum, 37, 58
    annealing, 38, 39
    austenizing, 38-39
    control, 38
    nondestructive testing, 38
    normalizing, 38, 39
    precipitation, 37
    solution, 37
    spheroidizing, 38, 39
    steel, 38
Heating
    electrical resistance, 110-11
Hexagonal close-packed lattice, 32, 33
Honeycomb, 162-63
Hot pressing, 137
Hot rolling, 128-29
    continuous, 129
Hot working, 125-26

Inductive-repulsive forming, 145
Ingots, 124, 125, 128, 129
Injection molding, 159
Inspection, 12, 13, 183-97
    control charts, 186-87
    final, 185
    first-piece, 184
    in-process, 184
    nondestructive testing, 185
    receiving, 184
    sampling, 185
Inspection equipment, 190-91
Inspection reliability, 7
    nondestructive testing, 7
Interchangeability, 10
Investment casting, 92
Iron, 37-38, 42
    body-centered cubic, 124, 125
Izod test, 25

Knoop numbers, 28

Lacquers, 118
Laminates, 162-63
Laser, 167
Lay, 154, 196
Liquid honing, 175
Loading systems, 17
    bending loads, 18
    load compression, 17
    load reversal, 17
    tensile load, 17
Lost wax process, 92
Low alloy AISI steels, 53
Low alloy structural steels, 51, 53

Machinability, 153-54
Machine tools, 151-53
    boring, 151
    broaches, 151
    drill press, 151
    grinders, 151
    planers, 151
    shapers, 151
    straight line, 151
    turning, 151
Machining, 147
    abrasive, 150-51
    boring, 151
    broaching, 151
    chip formation, 148
    chip types, 149
    cutting tools, 148, 150
    defects, 150
    drilling, 151
    finish, 154
    grinding, 151
    lapping, 151
    machinability, 153
    milling, 151
    numerical control, 154-56
    planing, 151
    shaping, 151
    smear, 150
    surface finishes, 147
    tool, 148
    tool motion, 148
    turning, 151
Macroporosity, 83, 125
Magnesium alloys, 64
Magnetostriction, 167
Manual welding, 108
Manufacturing, 73-78
    definition, 9
    history, 10
    markets, 74
    processes, 75-76
Martempering, 40

Martensite, 39, 101
Martensitic stainless steel, 53
Material failures, 3-6, 18, 28-29
    causes, 4-5
    definition, 4
    fatigue failure, 4, 23
    fracture, 4
    permanent deformation, 4
    progressive failure, 4
Material identification systems, 54-55
Material testing, 18
    destructive, 18
    direct, 18
    indirect, 18
    nondestructive, 18
    standardized tests, 19
Materials, 11-12
    atomic structure, 16
    chemical properties, 16
    mechanical properties, 16
    nonferrous, 124
    physical properties, 16
    processing properties, 16
    properties, 11, 15-29
Mean, 186
Measurement, 187
    air gages, 194
    attributes, 189
    comparators, 193-95
    contact pressure, 189
    dial indicators, 193
    dimensions, 187
    electrical gages, 193-94
    English system, 190
    error, 187
    fixed gages, 195-96
    gage blocks, 190
    gaging, 189
    human element, 189
    indicating gages, 193-95
    length standard, 190
    metric system, 190
    micrometer caliper, 191-92
    optical comparators, 194
    optical flats, 194
    parallax, 188
    pneumatic gages, 194
    sine bars, 192-93
    standards, 189
    steel rules, 191
    surface finish, 196
    temperature effects, 189
    tolerances, 187
    variation, 188-89
    Vernier caliper, 191, 192
Melting equipment, 94
    crucible furnaces, 94
    cupola, 94
    electric arc furnaces, 94
    induction furnaces, 94
    pot furnaces, 94
    reverbatory furnaces, 94
Metallic structures, 32-33
Metallizing, 179
Metallurgy
    powder, 135-36
    welding, 99-101
Metals, 135
    body-centered cubic, 136
    face-centered cubic, 136

Metals *(cont.)*
   failure, 123
   nonferrous, 57-65, 124
   processing, 75-76
   refractory, 138
   rolling, 128
   sheet, 141-46
   solidification, 80-82
Metric system, 190
Micrometer caliper, 191-92
Microporosity, 83
Microshrinkage, 83
Millwork, 128-33
Misruns, 85
Modulus of elasticity, 20
Mold, 80, 89-90
   dry sand, 90
   floor and pit, 90
   green sand, 89-90
   metal, 90
   shell, 90
Molding plastic, 157
   closed die, 157
   compression, 157
   injection, 157
   transfer, 157
Monomer, 67

Nickel, 62
Nickel alloys, 62, 65
   composition, 63
   corrosion resistance, 62
   properties, 63
   uses, 63
Nitriding, 138
Nondestructive testing, 5, 18, 123, 128, 185
   acoustic emission monitoring, 35, 44
   for brazing, 98
   chemical spot tests, 60
   for corrosion detection, 44
   definition, 1
   demonstration programs, 8
   in design, 75
   eddy current tests, 25, 38, 39, 44, 58, 60, 126, 132, 133, 142, 172, 179, 189
   electrical conductivity testing, 60
   fluoroscopic, 132, 138
   of forgings, 133-34
   of honeycomb, 163
   information sources, 2
   magnetic particle, 39, 44, 53, 134
   magnetic rubber, 44
   for maintenance, 126
   in manufacturing, 78
   neutron radiography, 44
   penetrant, 39, 44, 64, 93, 134, 136, 142, 178, 179, 181
   of pipe, 132
   of pipe and tubing, 130
   powdered metal purpose, 136
   purpose of, 6
   qualification and certification of personnel, 2
   radiation thickness gage, 130
   radiography, 33, 44, 64, 82, 89, 93, 124, 132, 135, 138
   of seamless tubing, 132
   of sheet metal, 142
   spectrographic analysis, 44
   supervisory personnel, 2
   symbols, 120
   thermo-electric, 38
   thickness control and measurement, 142
   ultrasonic testing, 33, 44, 53, 82, 124, 125, 132, 134, 142, 167
   ultrasonic thickness gage, 130
   for weldments, 97
Nonferrous alloys, 64-65
   corrosion-resistant, 64-65
   heat-resistant, 64-65
Nonferrous metals, 57-65
   alloying with iron, 58
   applications, 66-67
   characteristics, 66-67
   corrosion resistance, 58
   zinc, 58
Normal distribution, 186
Normalizing, 38, 39
Notched bar testing, 24
   Charpy test, 24-25
   Izod test, 25
Numerical control, 154-56
Nylon, 68

Open-hearth furnace, 49
Open-hearth steel, 48-49
Operating characteristic curve, 185
Oxidation, 126, 129
Oxyacetylene welding, 106

Parallax, 188
Pattern, 80
Patternmaker's shrinkage, 84
Pearlite, 39, 101
Percussive welding, 107
Permanent deformation, 143
Permeability, 89
Phase changes, 37
Phosphate coatings, 181
Pickling, 129, 174
Piercing mill, 132
Pig, 47
Pig iron, 47
Pipe, 130
   roll welding, 130-31
   spiral-welded, 131
Plasma-arc welding, 11
Plaster mold casting, 93
Plastic deformation, 34-35, 123
   cold work, 35
   fibering, 35
   rotational, 35
   slip, 34-35
   twinning, 35
Plastic flow, 19, 34, 121-26, 137, 142, 149
Plastic processing, 157-61
   casting, 159
   closed die molding, 157-59
   compression molding, 157, 158
   extrusion, 159
   injection molding, 158-59
   postforming, 160
   reinforced plastic molding, 159-60
   transfer molding, 158-59
Plastics, 65, 67-71
   cellulose, 68
   characteristics, 69-71
   definition, 67
   films, 159
   properties, 68, 160-61
   reinforced, 159
   synthetic, 68
   thermoplastic, 67, 69-70, 158, 159, 160, 161
   thermosetting, 67-68, 158, 159
Plate, 129
Plating
   electroplating, 180
   hot dip, 179-80
Polishing, 176
Polymerization, 67
Polymers, 67
Porcelain, 179
Porosity, 83
Portland cement, 163
Powder metallurgy, 135-36
   cemented carbides, 138
   compacting, 137
   composite materials, 139
   density, 137
   heat treatments, 138
   impregnation, 138
   machinability, 138
   postsintering, 138
   pressing, 136-37
   refractory, 138
   sintered bearings, 138-39
   sintering, 136, 137-38
Precipitation, 37
Precipitation hardening, 138
Pressworking
   bending, 142, 143-44
   deformation, 142
   drawing, 142, 144-45
   forming, 144
   shearing, 142
   sheet metal, 141-46
   stretch forming, 145
Process annealing, 36
Processes, 12
   adhesive bonding, 161
   arc cutting, 169
   bending, 143-44
   carburizing, 172
   casting, 79-94
   chemical milling, 166-67
   cleaning, 173
   cold finishing, 130
   deformation, 121, 123, 148
   drawing, 144-45
   electrical discharge machining, 164
   electrochemical machinery, 166
   electroforming, 168
   electromagnetic forming, 146
   explosive forming, 145
   extrusion, 132-33
   flame hardening, 173
   forging, 133-39
   forming, 143
   friction sawing, 169
   hot rolling, 128
   machining, 147-48
   plastic molding, 157
   powder metallurgy, 135-36
   pressworking, 141-46
   roll forming, 144
   shape-changing, 76-78

Processes *(cont.)*
   shearing, 142-43
   spinning, 145
   stretch forming, 144
   surface finishing, 171
   torch cutting, 168-69
   ultrasonic machinery, 167
   welding, 95-103
Proof testing, 6, 7
Properties
   mechanical, 125-26
Punch, 142

Quality control, 183

Recovery, 35
Recrystallization, 35-37, 100, 116, 122, 124, 125, 126, 128, 137, 138, 144
   theory, 36
Recrystallization temperatures, 36
Recrystallize, 137
Reinforced plastic molding, 159-60
   autoclave molding, 140
   compression molding, 160
   contact layup, 160
   vacuum bag molding, 160
Resilience, 21
Risers, 86
Rockwell test, 27-28
Roll forming, 144
Rotational deformation, 35
Rubber, 163
Runners, 80

S-N curve, 23
Sacrificial metals, 41
Safety, factors of, 5, 6, 28-29
Sand, 86
Sand compaction, 88-89
Sand molding, 86-90
   green sand, 87
   procedure, 86-87
Scarfing, 129, 168
Seamless tubing, 132
Season cracking, 43
Segregation, 82
   dendritic, 82
   ingot-type, 82, 125
Sheet, 129, 130
Sheet metal, 145
Shotpeening, 175
Shrinkage, 82-84
Shrinkage cavities, 125
Sintering, 136, 137-38
Skelp, 130
Slabs, 128
Slag, 46, 47-48, 108, 125
Slip deformation, 34-35
Slitting, 142, 143
Soldering, 98-99
Solid-state bonding, 113
Solidification, 33, 84-85
   directional, 84
   progressive, 84
Solidification of metals, 80-82
Solidification shrinkage, 83
Solution heat treatment, 37
Space lattice, 32
Spheroidizing, 38, 39
Spinning, 38, 39

Sprue, 80
Stainless steel, 51, 53-54, 65
   austenitic, 53, 124
   composition, 55
   corrosion resistance, 53
   ferritic, 53
   martensitic, 53
   properties, 55
   uses, 55
Standard deviation, 186
States of matter, 32
Statistical analyses
   detection probability, 7
   level of confidence, 7
Steel, 46, 47-54
   AISI basic classification, 55
   alloy, 51
   basic oxygen, 49
   bessemer, 48, 49
   carbon content, 47
   carburizing, 172
   casehardening, 172
   cast, 54
   composition, 51
   corrosion resistance, 51
   crucible, 48
   electric furnace, 49
   grain size, 51
   hardenability, 51
   hardening, 39-40
   heat treatment, 38
   high steel, 51
   low carbon, 50-51
   medium carbon, 51
   open-hearth, 48
   pickling, 40
   tool and die, 54
   toughness, 51
   weldability, 51
Steel making, 48-50
Steel sheet
   cold-finished, 125
Strain hardened, 122
Strain hardening, 125
Strain rate, 122
Stress, 17
   compressive, 18, 125, 128
   normal, 17
   shear, 17, 132, 142, 149
   tensile, 17
   unit, 17
Stress-strain diagram, 19
Stress corrosion, 43
Stress risers, 4
Stress rupture strength, 24
Strip, 129, 130
Stud welding, 107
Sublimation, 32
Submerged arc welding, 109
Superheat, 83, 85
Surface finish, 125, 196-97
   flaws, 196
   measurement, 196
   roughness, 196
   waviness, 196
Surface finishing, 171-97
   casehardening, 172
Surface specification, 197
   symbols, 197
Swaging, rotary, 134

Symbols
   nondestructive testing, 120
   welding, 114-15
Synthetic plastics, 68

Tempering, 40
Tensile impact test, 25
Tensile testing, 19
   specimens, 19
Testing
   bend, 25
   compression, 21
   creep, 24
   fatigue, 23-24
   hardness, 25-28
   notched bar, 24-25
   shear, 23
   transverse rupture, 22
Thermoplastic plastics, 67, 69-70
Thermosetting plastics, 67-68, 71
Tolerances, 188
   basic dimensions, 188
   understood tolerances, 88
Tool and die steels, 54
   chromium, 54
   manganese, 54
Torch cutting, 168-69
Toughness, 21
Transfer molding, 158-59
Transformation, 125-25
Transformation temperature, 124
Transition temperature, 4
Transverse temperature, 4
Transverse rupture testing, 22
Triple point, 32
True stress-true strain, 21
Tube, 130
Tube and pipe making, 130-32
Tubing
   extruded, 132
   resistance welded, 131
   seamless, 132
Tukon microhardness tester, 28
Turning, 151
Twinning deformation, 35

Ultimate strength, 20, 21
Ultrasonic machining, 167
   finishes, 167
   transducer, 167
Ultrasonic welding, 112
Upsetting, 135

Vacuum metallizing, 179
Varnish, 177
Vernier, 192
Vickers test, 28
Vitreous enamels, 179

Waviness, 154
Weld defects, 116-20
Weld joints, 113-15
   butt joints, 114
   corner joints, 114
   edge joints, 114
   lap joints, 114
   tee joints, 114
Weldability, 115-16
Welding, 95-103, 148
   angular distortion, 102

Welding *(cont.)*
  automatic, 109-10
  base material, 100-101
  defined, 95
  diffusion, 113
  distortions and stresses, 101-3
  electric arc, 106-7
  electrodes, 107-8
  electron-beam, 11
  electroslag, 112
  explosion, 112-13
  filler, 97, 100, 108
  forge, 106
  friction, 112
  fusion, 96-96, 117
  gas metal-arc, 108
  gas shielding, 107
  gas tungsten-arc, 108
  gas tungsten wire, 109
  lateral distortion, 102
  manual, 108
  metallurgy, 99-101
  oxyacetylene, 106
  percussive, 107
  plasma-arc, 11
  post treatment, 101
  preheating, 101
  pressure, 97
  projection, 110
  resistance, 110, 131
  restraints, 102
  reverse polarity, 107
  seam, 110
  shielding gases, 108
  spot, 110
  straight polarity, 107
  stud, 107
  submerged arc, 109, 131
  symbols, 114-15
  ultrasonic, 112
Welding bell, 130
Welding defects, 116-20
  corrosion resistance, 101, 116
  cracking, 101
  cracks, 119
  crater cracks, 119
  dimensions, 117
  dissolved and entrapped gases 99-106
  distortion, 116
  double vee, 117
  fillet, 117
  fusion zone, 101
  grain structure, 100
  heat-affected zone, 119
  inadequate joint penetration, 118
  inclusions, 100-118
  incomplete fusion, 118
  joints, 113-15
  lap, 131
  porosity, 118
  postcracking, 103
  postheating, 103
  profile, 117
  residual stresses, 102
  slag and oxides, 100
  surface irregularities, 119
  undercut, 118-19
  warping, 117
Weldment, 95
Wire brushing, 175
Work hardened, 122
Work hardening, 34, 122
Wrought aluminum, 59
Wrought iron, 47-48

X (mean), 186

Yield point, 20
Yield strength, 20
Young's modulus, 20

Zinc alloys, 64